MODIFIED FIBERS WITH MEDICAL
AND SPECIALTY APPLICATIONS

Modified Fibers with Medical and Specialty Applications

Edited by

J. VINCENT EDWARDS

Southern Regional Research Center,
New Orleans, LA, U.S.A.

GISELA BUSCHLE-DILLER

Auburn University
Aburn, AL, U.S.A.

and

STEVEN C. GOHEEN

Pacific Northwest National Laboratory,
Richland, WA, U.S.A.

 Springer

A C.I.P. Catalogue record for this book is available from the Library of Congress.

ISBN-10 1-4020-3793-7 (HB)
ISBN-13 978-1-4020-3793-1 (HB)
ISBN-10 1-4020-3794-5 (e-book)
ISBN-13 978-1-4020-3794-8 (e-book)

Published by Springer,
P.O. Box 17, 3300 AA Dordrecht, The Netherlands.

www.springer.com

Printed on acid-free paper

Printed in the Netherlands.

Table of contents

1. THE FUTURE OF MODIFIED FIBERS   1
    J. Vincent Edwards, Steven C. Goheen, and Gisela Buschle-Diller

2. FUTURE STRUCTURE AND PROPERTIES OF
    MECHANISM-BASED WOUND DRESSINGS   11
    J. Vincent Edwards

3. BEHAVIOR OF CELLS CULTURED ON CUPROPHAN   35
    N. Faucheux, J. L. Duval, J. Gekas, M. Dufresne,
    R. Warocquier, and M. D. Nagel

4. COTTON AND PROTEIN INTERACTIONS   49
    Steven C. Goheen, J. Vincent Edwards, Alfred Rayburn,
    Kari Gaither, and Nathan Castro

5. ELECTROSPUN NANOFIBERS FROM BIOPOLYMERS
    AND THEIR BIOMEDICAL APPLICATIONS   67
    Gisela Buschle-Diller, Andrew Hawkins, and Jared Cooper

6. HALAMINE CHEMISTRY AND ITS APPLICATIONS
    IN BIOCIDAL TEXTILES AND POLYMERS   81
    Gang Sun and S. D. Worley

7. MODIFICATION OF POLYESTER FOR MEDICAL
    USES   91
    Martin Bide, Matthew Phaneuf, Philip Brown,
    Geraldine McGonigle, and Frank LoGerfo

8. BIOLOGICAL ACTIVITY OF OXIDIZED
    POLYSACCHARIDES   125
    Ioan I. Negulescu and Constantin V. Uglea

9. BIOLOGICAL ADHESIVES   145
    José María García Páez and Eduardo Jorge-Herrero

10. SURFACE MODIFICATION OF CELLULOSE FIBERS
    WITH HYDROLASES AND KINASES   159
    Tzanko Tzanov and Artur Cavaco-Paulo

11. ENZYMES FOR POLYMER SURFACE
    MODIFICATION   181
    G. Fischer-Colbrie, S. Heumann, and G. Guebitz

12. ENZYMATIC MODIFICATION OF FIBERS FOR
    TEXTILE AND FOREST PRODUCTS INDUSTRIES   191
    William Kenealy, Gisela Buschle-Diller, and Xuehong Ren

13. THE ATTRACTION OF MAGNETICALLY
    SUSCEPTIBLE PAPER   209
    Douglas G. Mancosky and Lucian A. Lucia

14. FIBER MODIFICATION *VIA* DIELECTRIC-BARRIER
    DISCHARGE: Theory and practical applications to
    lignocellulosic fibers   215
    L. C. Vander Wielen and A. J. Ragauskas

INDEX   231

Preface

The initial impetus for this book on fibers originated from a weeklong symposium where scientists of a variety of walks met to discuss their work on fibers with medical and specialty applications. Seeing the benefits of sharing information across disparate fields and disciplines of science we realized the potential for cross-fertilization of ideas between different area of fiber science. Thus, represented here are a variety of potential product lines under the cover of a single book, which for the imaginative scientist we hope will lead to some new food for thought. The fields of medical and specialty fibers include a wide array of natural and synthetic textiles, medical devices, and specialty paper and wood products. Research in these areas has become more interesting to scientists who are seeking to strike out in new directions based on an impulse to create new products that meet the unmet needs of rapidly growing fiber markets in wound care, prosthetic, and cellulosic arenas. It is hoped that providing new concepts and approaches to working with different types of fibrous materials will give the reader some pulse of the current climate and research opportunities of medical and specialty fibers. Breakthroughs into a better understanding of wound healing, biomaterial design, fiber surface chemistry and bio- and nanotechnologies are currently providing the impetus to create the fiber products of the future. The editors feel that a book of this type would be remiss without discussions of the impact interdisciplinary scientific pursuits are having on fiber design. With that in mind we have treaded lightly on reviewing traditional areas that have been the basis of past books on fibers science, and provide papers giving emphasis to chemically, biologically, and material science oriented readers. Included here are papers by featured authors who have or are currently developing new fiber products in wound dressing, hygienic and cellulosic products. Medical textiles provide the foundation for current medical technology products of the future. Subjects on fiber design and modification dealing with non-implantable, implantable, and extracoporeal materials, are provided for in the first nine chapters. The interdisciplinary nature of textile fiber science includes areas from physics to biology; and the boundaries between seem to be growing fainter as new fibers modifications are being developed. It is with this in mind that the final chapters 10–14 are presented giving new insights to areas of fiber and enzyme and surface physics and issues that present new research concepts on the molecular engineering and physics of cellulosic fibers.

Chapter 1

The Future of Modified Fibers

J. Vincent Edwards[1], Steven C. Goheen[2], and Gisela Buschle-Diller[3]

[1] *USDA-ARS, Southern Regional Research Center, 1100 Robert E. Lee Blvd., New Orleans, LA 70124, U.S.A.*
[2] *Battelle Northwest, Richland, Washington 99352, U.S.A.*
[3] *Textile Engineering Department, Auburn University, AL 36849, U.S.A.*

The future of fiber technology for medical and specialty applications depends largely on the future needs of our civilization. It has been said that "unmet needs drive the funding that sparks ideas". In this regard recent emphasis on United States homeland security has encouraged new biofiber research, resulting in the development of anti-bacterial fibers for producing clothing and filters to eliminate pathogens and enzyme-linked fibers to facilitate decontamination of nerve toxins from human skin [1]. Magnetic fibers may also have future security applications including fiber-based detectors for individual and material recognition. Interest in smart and interactive textiles is increasing with a projected average annual growth rate of 36% by 2009 [2]. More specific markets including medical textiles and enzymes will grow even more rapidly. Among the medical textiles are interactive wound dressings, implantable grafts, smart hygienic materials, and dialysis tubing. Some of the medical and specialty fibers inclusive of these types of product areas are discussed in this book. A recent review of the surface modification of fibers as therapeutic and diagnostic systems relevant to some of these new product areas has appeared and Gupta reviewed current technology for medical textile structures [3] with focus on woven medical textile materials.

The design of new fibers for use in healthcare textiles has increased rapidly over the past quarter of a century. Innovations in fiber design have led to improvements in the four major areas of medical textiles: non-implantable, implantable, extracorporeal, and hygienic products. The use of natural fibers in

J. V. Edwards et al. (eds.), Modified Fibers with Medical and Specialty Applications, 1–9.

medical applications spans to ancient times. Although wood seems an unlikely material for a medical textile, some of the earliest documented evidence of the use of natural fibers as prosthetics is from the use of wooden dentures in early civilizations [4]. Anecdotal folklore also suggests that President George Washington wore similar prosthetics; however his dentures were probably constructed of ivory [5]. It is notable that wood is still employed in splints to stabilize fractures [6]. Natural fibers are readily available and easily produced owning to their remarkable molecular structure that affords a bioactive matrix for design of more biocompatible and intelligent materials. The nanostructure of natural fibers is complex and organized in motifs that cannot be easily duplicated. Synthetic fibers typically do not have the same multilevel structure as native materials. On the other hand, specific material properties including the modulus of elasticity, tensile strength, and hardness are largely fixed parameters for a natural fiber but have been more manageable within synthetic fiber design. The molecular conformation native to natural fibers is often key to interactions with blood and organ cells, proteins, and cell receptors, which are currently being studied for a better understanding to improve medical textiles. The native conformation or periodicity of structural components in native fibers such as collagen and cellulose offers unique and beneficial properties for biomedical applications. An extension of the bioactive conformation property in fibers to rationally designed fibers that would inhibit enzymes or trigger a cell receptor is a premise of current research.

The first nine chapters of this book present work going on in the research and development of biomedical products from these four traditional areas of medical textiles.

Non-implantable textiles are applied externally. They include dressings and bandages used in wound and orthopedic care, bedpads, sheets, diapers, and protective clothing such as patient and medical personnel gowns, gloves, face masks, and related items. Non-implantable wound dressings are largely exposed to the skin and wound fluid as well as subcutaneous cells [7]. Chapters 2 and 4 both discuss recent results of work in an area of mechanism-based non-implantable fibers that address a current need to enhance wound healing by redressing the molecular imbalance of the chronic wound. Wound healing and material science are shaping new views on how dressings are being improved and expected to develop. The implications of mechanism-based dressings employing the concepts of contemporary wound bed preparation and wound healing science for future chronic wound dressings are drawn from the current state of the science. The two natural fibers collagen and cellulose play an important role in new wound dressing designs. The most common application for collagen in dermatology is tissue augmentation and wound healing [8]. An example of collagens role in non-implantable materials is evident in interactive wound dressings, which have a mechanism-based mode of action and employ either

Crystallite

Unit Cell

Cellulose Chain

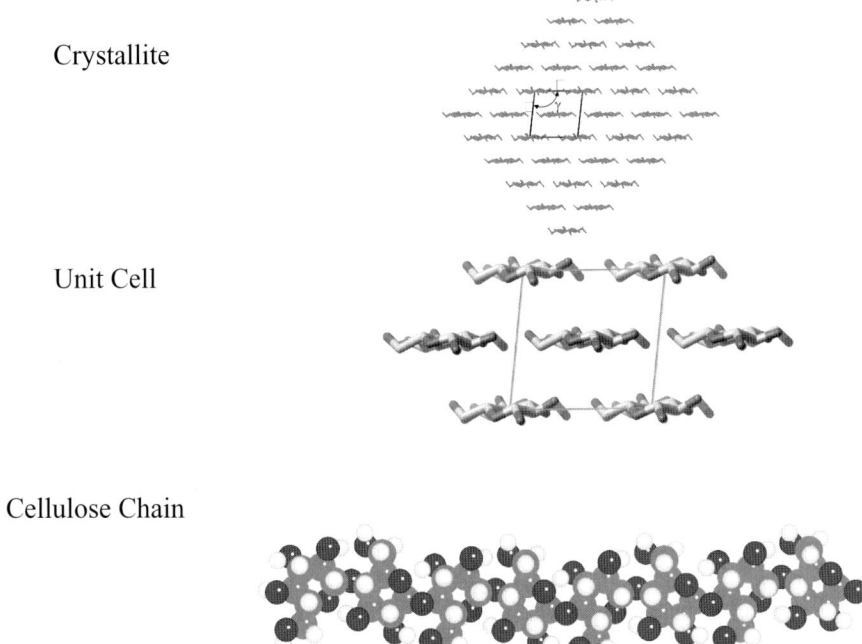

Figure 1.1. A portrayal of the levels of structure of cellulose (structures are provided courtesy of Dr. Alfred D. French). The cellulose chain, which is an unbranched chain of glucose residues with ß-(1–4) linkages. The second level of structure is the unit cell, which is shown here as a cross-section of cellulose chains. The unit cell is the smallest piece of a crystal that can be repeated in the x, y, and z directions to generate an entire crystal. Here, it consists of two cellobiose units. One is located at the corners of the unit cell and another at the center. Although there are chains at each corner, only one-fourth of each is inside the unit cell for a total of one corner chain. This crystallite contains 36 chains and is thought to correspond to an elementary fibril for higher plant secondary walls. Its atomic positions, like those in the unit cell, is based on the structure of cellulose that was reported in Nishiyama, Y.; Langan, P.; Chanzy, H. Crystal structure and hydrogen-bonding system in cellulose Iß from synchrotron X-ray and neutron fiber diffraction. *J. Am. Chem. Soc.* **2002**, *124*, 9074–9082.

a native or electrospun form of collagen fibers to stimulate cell growth and to augment soft tissue repair.

Collagen is a key component in several different tissues, and though the fibrous form of the protein is varied it fulfills the requirements of an important structural component of both non-implantable and implantable materials. Collagen possesses multiple levels of structure (Figure 1.1), which are interesting to contemplate for its role in a variety of biocompatible materials as viewed. Collagen has a repeating amino acid sequence. Two out of three of these sequences are identical (alpha-1) left-handed helices with a pitch of 9.5 Å. The third is a nearly identical (alpha-2) chain with the same left-handed pitch. These three strands of amino acids are bound together in a right-handed triple helix with a pitch of 104 Å. These helices are coupled by hydrogen bonds between the HN group of glycine in one chain and O=C groups of an adjacent amino acid. Each super helix is about 1000 residues long, and these residues are staggered to form 668 Å repeating units at the higher structural level, the microfibril. Microfibrils are further organized at several levels resulting in the final structure of collagen.

Other natural fibers such as elastin, silk, and wool, which are also proteinaceous are as complex and unique as cellulose and collagen. Some researchers have examined ways to modify wool [9, 10] and silk [9, 11] to enhance their bacterial resistance. The work with these fibers has been expanded to include other natural fibers and the enhancement of anti-fungal properties [12]. Silk is also commonly used for sutures although may not be as effective as other tissue sealing methods when underivatized [13] and may some day be used to augment bone repair [14, 15]. Genetically engineered forms of elastin have been used for cartilage tissue repair [16]. Closely related research areas address the ability of natural or synthetic fibers to either resist microbe adhesion [17] or produce anti-microbial fabrics from other fibers.

Cellulose is similar in its structural complexity to collagen. However, cellulose is composed of carbohydrate residues. Differences between cotton and wood cellulose, for example, are significant at the macromolecular level, but the molecular sequences are similar. In the cotton fiber, many levels of organization have been discovered based on the arrangement of the crystalline microfibrils that are ordered in multilayer structures. Figure 1.2 demonstrates an analog of progressing from the smallest unit that is the cellulose molecule in-

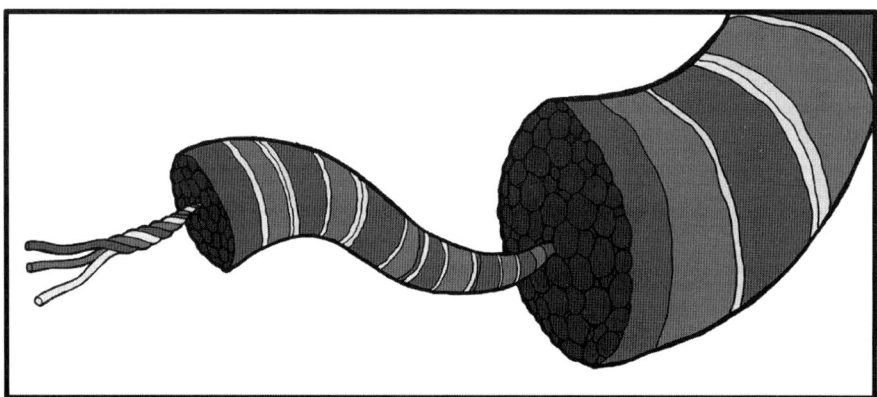

Figure 1.2. A simplified illustration representing the three major levels of structure of collagen fibers: Triple helical collagen (3000 Å by 16 Å) molecules are packed into collagen microfibrils that are assembled into the native collagen fiber

visible to light microscopes to the cotton fiber visible to the naked eye. Chapter 4 examines blood proteins, their adsorption to cotton, and their potential role in wound healing. Much of the concern about modified fiber performance for medical applications involves the interface between the fiber and its immediate environment. In Chapter 4, Goheen et al. present an approach to understanding the interaction of the blood protein albumin with a modified cotton wound dressing fiber and an enzyme that takes up destructive residence in chronic wounds. In Chapter 6, Sun and Worley present current product-oriented work on a type of non-implantable hygienic textile with biocidal activity, in which they attach halamines to the surface of cotton and cotton/polyester fibers. This work is an important chapter in the development of regenerable anti-microbial fabrics and represents a growing effort to control microbes in hospital textiles and protective fabrics. Modified cellulose has also been used to generate microcapsules to deliver pharmaceuticals [18]. There has been recent research on the use of modified cellulose derivatives to create ultra thin coatings on biomaterials [19]. Regioselectively derivatized cellulose has also been explored for its anti-coagulant activity, which is another example of bioactive fibers from biopolymers. In Chapter 8, Negulescu et al. further discuss the bioactive polymer idea from a drug discovery paradigm and give examples from their own work of biologically active polysaccharide polymers from plants. Indeed, polysaccharide fibers offer interesting possibilities for drug discovery from both rational design and combinatorial motifs.

In Chapter 7, Bide et al. review the medical uses of polyester fibers, which along with polytetrafluoroethylene predominate the market of vascular grafts. Implantable fibers are placed *in vivo* for wound closure or replacement surgery. Factors in determining the biocompatibility of a textile include biodegradability, toxicity, fiber size, porosity, and tissue encapsulation. Implantable medical textile product groups that are currently being researched and developed are arterial grafts, surgical sutures, stents, and ligaments. An important area of research is concerned with improving the fabric failure of conventional grafts within the harsh hemodynamic milieu especially when coupled to stents [20]. Vascular grafts have been used for over 40 years to replace diseased or damaged arteries. Implants are also exposed to several different types of tissues, depending on the location of the implant. Much of the current interest in fiber biocompatibility with fluids and tissues reverts to the compatibility between the implant (or wound dressing) and the proteins in the immediate environment. Protein binding to implant materials has been the subject of a large body of literature over several decades. To summarize this body of literature on protein/material binding the statement *"water soluble proteins tend to resist binding to highly hydrophilic surfaces"* conceptualizes the primary issue. This property of protein/material binding exists because water forms a partially impenetrable layer between the protein and the surface. However, hydrophilic surfaces are not necessarily more biocompatible than hydrophobic surfaces. In this regard, it is still not entirely clear whether blood coagulation and tissue rejection can be predicted based on simple surface parameters as surface tension determinations.

In Chapter 3, Faucheux et al. examine cell behavior and some key cellular mechanisms of proliferation and programed cell death in the presence of serum on a Cuprophan-modified surface. Extracorporeal fibers are those used in mechanical organs such as hemodialysers, artificial livers, and mechanical lungs. Historically regenerated cellulose fibers in the form of cellophane have been utilized to retain waste products from blood. Cuprophan, a cellulosic membrane, has been the material of choice due to the selective removal of urea and creatinine while retaining nutritive molecules such as vitamin B12 in the bloodstream. Other medical applications of modified cellulose include hemodialysis membranes (vitamin E modified cellulose [21]) and cellulose diacetate membranes [22]. A more thorough understanding of how the surface properties of extracorporeal fibers which are in contact with blood effect cells in the presence of blood proteins will improve our understanding of improved fiber design and modification.

In Chapter 9, García Páez and Jorge-Herrero introduce work on the uses and preparation of biological adhesives, which is vital to tissue engineering. Tissue engineering is a discipline of biotechnology that creates biological scaffolds for the stimulation of cell growth, differentiation, viability, and the development of functional human tissue. Some of the first commercial tissue engineering products, which focused on skin replacement, will be covered in this chapter. However, technologies are under development to address the pathology of virtually every tissue and organ system. A promising area of tissue engineering is the growing research on fibrin sealants and tissue adhesives for surgical use, acceleration of wound healing, and regeneration of damaged tissue.

Tissue engineering also employs both natural and synthetic polymers electrospun into fibers. These electrospun fibers include collagen, elastin, gelatin, fibrinogen, polyglycolic acid, polylactic acid, polycapronic acid, and others. It has been said that this is the decade of nanoengineered materials, and in the area of medical science product potential it is virtually limitless. In Chapter 5, Buschle-Diller et al. highlight some of the principles of electrospun nanofibers and biomedical fibers of interest.

Chapters 10–12 present emerging concepts on enzyme applications to both natural and synthetic fibers. The inclusion of these three chapters on specialty applications alongside chapters for medical fibers is timely with the current interest of applying biotechnology to fibers. At a molecular level, there are close similarities between the biological modification of a fiber with an enzyme and the biological activity of a modified fiber through inhibition or promotion of enzyme activity. At this chemical/biological interface of subject areas, interest often becomes interdisciplinary and new ideas may be spawned. It is also very evident that the scientific community is now turning to enzymes in an effort to make our world more renewable and sustainable. Although enzymes have been used in textile processing for many years, it is only in the last 20 years that growing interest has been given to using a variety of enzymes for textile and fiber applications. Thus, in Chapter 10, Tzanov and Cavaco-Paulo reveal new approaches to modifying cellulose fibers with enzymes applied to the two long-studied problems of fabric crease-resistance and flame retardant finishing. The approach of surface modifying a synthetic fiber is taken up by Fischer-Colbrie et al. in Chapter 11 in the context of hydrolytic and oxidative enzymes, and their application to the many fiber surfaces that are structural components of the modern world. Finally, Kenealy in Chapter 12 extend the coverage to enzymatic modification of fibers in textile and forest products. In the closing two chapters of the book, we have come full circle from wooden dentures in ancient civilizations to the treatment of lignocellulose-containing wood and paper with cold plasmas (Chapter 13) and magnetic susceptibility properties (Chapter 14), respectively. These two chapters also turn our attention further to new technologies and green chemistries that open up promising ways of modifying lignocellulosic fibers.

Some imaginative questions that one might pose as these chapters are being read are, how will fiber technology evolve? We already have numerous military and civilian benefits from fiber development. We have clothes that selectively repel liquid water while allowing the penetration of water vapor. Will biotechnology help us design fibers or polymers to withstand intense radiation while maintaining their integrity? Will we discover that the nanostructure of natural fibers is ideal for implant biocompatibility, thereby opening the door for more successful developments of synthetic replacement organs? How interactive can we expect textile fibers of the future to be? Will we learn from natural fibers how to design synthetic fibers for better control of surface and bulk properties? We leave it to the reader to pose further imaginative questions regarding the future of modified fibers.

The technologies mentioned here are rapidly developing, but it is the editors' belief that the chapters included in this book offer current information that will form a part of the basis of future discoveries in modified fiber technology.

References

1. Grimsley, J. K.; Singh, W. P.; Wild, F. R.; Giletto, A. A novel, enzyme-based method for the wound surface removal and decontamination of organophosphorus nerve agents. In: Edwards, J. V.; Vigo, T. L. (Eds.) *Bioactive Fibers and Polymers.* American Chemical Society, Washington, DC, **2001**, 35–49.

2. http://www.bccresearch.com/editors/RGB-309.html.
3. Gupta, B. S. Medical textile structures: An overview. *Med. Plast. Biomater.* **1998**, *5*(1), 16–30.
4. Engelmeier, R. L. The history and development of posterior denture teeth—Introduction, part I. *J. Prosthodont.* **2003**, *12*(3), 219–226.
5. Glover, B. George Washington, a dental victim. *Riversdale Lett.* **1998**, *16*(62).
6. Honsik, K.; Boyd, A.; Rubin, A. L. Sideline splinting, bracing, and casting of extremity injuries. *Curr. Sci.* **2003**, *2*, 147–154.
7. Wollina, U.; Heide, M.; Muller-Litz, W.; Obenauf, D.; Ash, J. Functional textiles in prevention of chronic wounds, wound healing and tissue engineering. Textiles and the Skin, Karger, Basel, Switzerland, **2003**, 82–97.
8. Ruszczak, Z.; Schwartz, R. A. Collagen uses in dermatology—An update. *Dermatology* **1999**, *199*(4), 285–289.
9. Abel, T.; Cohen, J. I.; Escalera, J.; Engel, R.; Filshtinskaya, M.; Fincher, R.; Melkonian, A.; Melkonian, K. Preparation and investigation of antibacterial protein-based surfaces. *J. Text. Apparel Technol. Manage.* **2003**, *3*(2), 1–8.
10. Choi, H.-M.; Bide, M.; Phaneuf, M.; Quist, W.; LoGerfo, F. Dyeing of wool with antibiotics to develop novel infection resistance materials for extracorporeal end use. *J. Appl. Polym. Sci.* **2004**, *92*(5), 3343–3354.
11. Tsukada, M.; Katoh, H.; Wilson, D.; Shin, B.-S.; Arai, T.; Murakami, R.; Freddi, G. Production of antimicrobially active silk proteins by use of metal-containing dyestuffs. *J. Appl. Polym. Sci.* **2002**, *86*(5), 1181–1188.
12. Cohen, J. I.; Abel, T.; Burkett, D.; Engel, R.; Escalera, J.; Filshtinskaya, M.; Hatchett, T.; Leto, M.; Melgar, Y.; Melkonian, K. Polycations. 15. Polyammonium surfaces—A new approach to antifungal activity. *Lett. Drug Des. Discov.* **2004**, *1*(1), 88–90.
13. Giray, G. B.; Atasever, A.; Durgun, B.; Araz, K. Clinical and electron microscope comparison of silk stutres and *n*-butyl-cyanocrylate in human mucosa. *Aust. Dent. J.* **1997**, *42*(4), 255–258.
14. Sofia, S.; McCarthy, M. B.; Gronowicz, G.; Kaplan, D. L. Functionalized silk-based biomaterials for bone formation. *J. Biomed. Mater. Res.* **2000**, *54*(1), 139–148.
15. Chen, J.; Altman, G. H.; Karageorgiou, V.; Horan, R.; Collette, A.; Volloch, V.; Colabro, T.; Kaplan, D. L. Human bone marrow stromal cell and ligament fibroblast responses on RGD-modified silk fibers. *J. Biomed. Mater. Res. Part A* **2003**, *67A*(2), 559–570.
16. Knight, M. K.; Setton, L. A.; Chilkoti, A. Genetically engineered, enzymatically crosslinked elastin-like polypeptide gels for cartilage tissue repair. 2003 Summer Bioengineering Conference, Sonesta Beach Resort in Biscayne, Florida, **2003**.
17. Ingham, E.; Eady, E. A.; Holland, K. T.; Gowland, G. Effects of tampon materials on the in-vitro physiology of a toxic shock syndrome strain of *Staphylococcus aureus. J. Med. Microbiol.* **1985**, *20*(1), 87–95.
18. Dautzenberg, H.; Schuldt, U.; Grasnick, G.; Karle, P.; Muller, P.; Lohr, M.; Pelegrin, M.; Piechaczyk, M.; Rombs, K. V.; Gunzberg, W. H.; Salmons, B.; Saller, R. M. Development of cellulose sulfate-based polyelectrolyte complex microcapsules for medical applications. *Ann. N. Y. Acad. Sci.* **1999**, *875*, 46–63.
19. Baumann, H.; Richter, A.; Klemm, D.; Faust, V. Concepts for preparation of novel regioselective modified cellulose derivatives sulfated, aminated, carboxylated and acetylated for hemocompatible ultrathin coatings on biomaterials. *Macromol. Chem. Phys.* **2000**, *201*(15), 1950–1962.
20. Melbin, J.; Ho, P. C. Stress reduction by geometric compliance matching at vascular graft anastomoses. *Ann. Biomed. Eng.* **1997**, *25*, 874–881.

21. Sasaki, M.; Hosoya, N.; Saruhashi, M. Vitamin E modified cellulose membrane. *Artif. Organs* **2000**, *24*(10), 779–789.

22. Gastaldello, K.; Melot, C.; Kahn, R.-J.; Vanherweghem, J.-L.; Vincent, J.-L.; Tielemans, C. Comparison of cellulose diacetate and polysulfone membranes in the outcome of acute renal failure. A prospective randomized study. *Nephrol. Dial. Transplant.* **2000**, *15*, 224–230.

Chapter 2

FUTURE STRUCTURE AND PROPERTIES OF MECHANISM-BASED WOUND DRESSINGS

J. Vincent Edwards

USDA-ARS, Southern Regional Research Center, 1100 Robert E. Lee Blvd., New Orleans, LA 70124, U.S.A.

Abstract

The research and development of chronic wound dressings, which possess a mechanism-based mode of action, has entered a new level of understanding in recent years based on improved definition of the biochemical events associated with pathogenesis of the chronic wound. Recently, the molecular modes of action have been investigated for skin substitutes, interactive biomaterials, and some traditional material designs as balancing the biochemical events of inflammation in the chronic wound to improve healing. The interactive wound dressings have activities including up-regulation of growth factors and cytokines and down-regulation of destructive proteolysis. Carbohydrate-based wound dressings have received increased attention for their molecular interactive properties with chronic and burn wounds. Traditionally, the use of carbohydrate-based wound dressings including cotton, xerogels, charcoal cloth, alginates, chitosan, and hydrogels have afforded properties such as absorbency, ease of application and removal, bacterial protection, fluid balance, occlusion, and elasticity. Recent efforts in our lab have been underway to design carbohydrate dressings that are interactive cotton dressings as an approach to regulating destructive proteolysis in the non-healing wound. Elastase is a serine protease that has been associated with a variety of inflammatory diseases and has been implicated as a destructive protease that impedes wound healing. The presence of elevated levels of elastase in non-healing wounds has been associated with the degradation of important growth factors and fibronectin necessary for wound healing. Focus will be given to the design, preparation, and assessment of a type of cotton-based interactive wound dressing designed to intervene in the pathophysiology of the chronic wound through protease sequestration.

J. V. Edwards et al. (eds.), Modified Fibers with Medical and Specialty Applications, 11–33.
© 2006 *Springer. Printed in the Netherlands.*

2.1 Historical characteristics of wound dressing fibers and wound healing

Through the ages, both vegetable and animal fibers have been applied to human wounds to stop bleeding, absorb exudate, alleviate pain and provide a protective barrier for the formation of new tissue. Some milestones of wound dressing development down through the ages are summarized in Figure 2.1. Early humankind employed many different materials from the natural surroundings including resin-treated cloth, leaves, and wool-based materials with a variety of substances including eggs and honey. Some of these ancient remedies were probably more than palliative treatments. For example, the antibacterial activity of honey in the treatment of wounds has been established [1], and honey is now being reconsidered as a dressing when antibiotic-resistant strains prevent successful antibiotic therapy. Recent studies suggest that honey may promote wound healing through stimulation of inflammatory cytokines from monocytic cells [2]. Leaves of *chromolaena odorata*, a weed found in crops in countries of the Southern Hemisphere, have been found to exert potent antioxidant effects

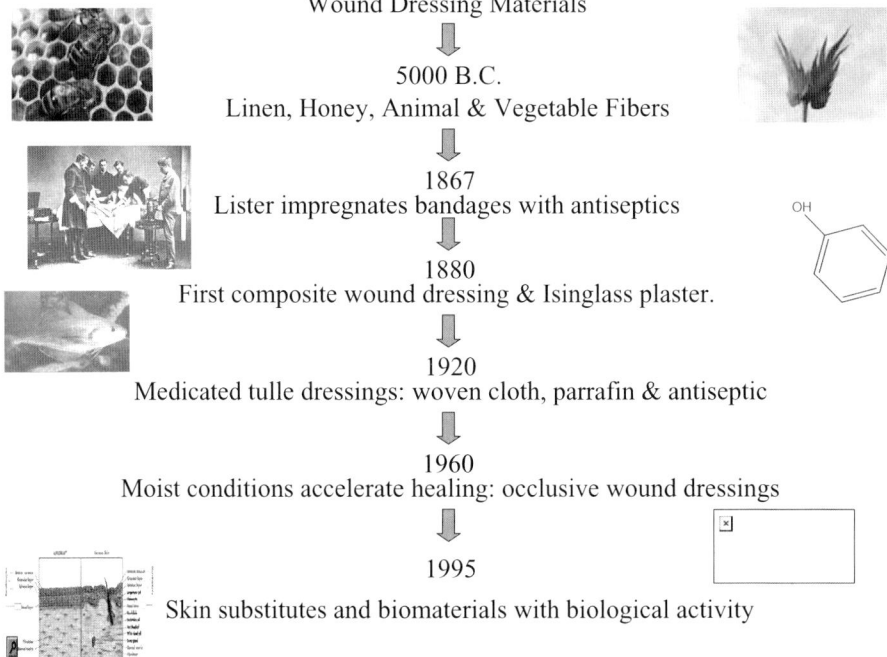

Wound Dressing Materials
⇩
5000 B.C.
Linen, Honey, Animal & Vegetable Fibers
⇩
1867
Lister impregnates bandages with antiseptics
⇩
1880
First composite wound dressing & Isinglass plaster.
⇩
1920
Medicated tulle dressings: woven cloth, parrafin & antiseptic
⇩
1960
Moist conditions accelerate healing: occlusive wound dressings
⇩
1995
Skin substitutes and biomaterials with biological activity

Figure 2.1. Timeline overview for historical developments in wound dressing materials. See *Wound Management and Dressings by Stephen Thomas, Pharmaceutical Press, London 1999* for an in-depth treatment of the development and history of wound dressings

that enable enhanced proliferation of human dermal fibroblasts and epidermal keratinocytes [3]. Wool-based dressings are also being rediscovered for their unique properties applicable to burn and chronic wounds [4]. Many ancient remedies for wound healing were contaminated with microorganisms, which increased the likelihood of infection. However, with the work of Lister in 1867, who impregnated bandages with carbolic acid, antiseptic treatment arose, and shortly thereafter Joseph Gamgee produced the first composite wound dressing as a cotton or viscose fiber medicated with iodine [5]. The first film dressings composed of Isinglassplaster were also introduced in the 18th century and were reported to be used with some improved success after skin grafting at that time. The first generation of medicated tulle dressings (see Contact Layer Dressings in Table 2.1) were introduced in the 1920s for treatment of burn wounds and were composed of an open weave cloth with soft paraffin and antiseptic. The finding that moist wounds heal faster than when desiccated and that collagen at the interface of the scab and dermis impedes epidermal cell movement prompted the development of occlusive dressings for wound management [6, 7]. In 1975, Rheinwald and Green [8] developed a method that made it possible to cultivate human keratinocytes so that a 1–2 cm^2 of keratinocyte cultured grafts in about 3 weeks. This work paved the way for the eventual development of skin substitutes and biomaterials with wound interactive properties and biological activity which has progressed from the mid 1990s through the present.

The science of wound healing has progressed rapidly over the past 30 years. An understanding of the progress of wound healing science, as seen by how the "future" of wound healing was viewed in the late sixties, can be seen from this quote taken from Christopher Textbook of Surgery:

Will the surgeon of 2000 AD encounter the same healing problems as the present-day surgeon? Let us hope not. Prudden's studies with cartilage have shown unquestioned stimulation of the healing process in a number of different healing situations; surely a purification and chemical dissection of this crude product can result in a sterile, more potent compound which can be used parenterally as well as locally. There may be no need to use such a compound in normal patients but it could prove invaluable in patients in whom impaired healing is to be expected [9].

If we fast forward 35 years from this quote it is found that the "potent compound" alluded to is today thought of as a variety of biologically potent protein families that play a central role in stimulating and regulating wound healing. These include growth factors, cytokines, and chemokines. Growth factors are mitogens that stimulate proliferation of wound cells. Growth factors are "messages for cells" with hormone-like potency, and as proteins they bind and activate specific cell receptors. They regulate gene expression, protein

Table 2.1.

Dressing and fiber type	Description	Properties	Indications	Contraindications
Thin films (bioclusive, tegaderm, opsite)	Semipermeable, polyurethane membrane with acrylic adhesive.	Permeable to water and oxygen providing a moist environment.	Minor burns, pressure areas, donor sites, postoperative wounds.	Deep cavity wounds, third-degree burns infected wounds.
Sheet hydrogels (clearsite, nu-gel, vigilon, geliperm, duoderm gel, intrasite gel, geliperm granulate)	Solid, non-adhesive gel sheets that consist of a network of cross-linked, hydrophilic polymers that can absorb large amounts of water without dissolving or losing its structural integrity. Thus they have a fixed shape.	Carrier for topical medications. Absorbs its own weight of wound exudate. Permeable to water vapor and oxygen, but not to water and bacteria. Wound visualization.	Light to moderately exudative wounds. Autolytic debridement of Stages II and III pressure sores.	Heavily exuding wounds.
Hydrocolloids (duoderm, comfeel)	Semipermeable polyurethane film in the form of solid wafers; contain hydroactive particles as sodium carboxymethyl cellulose that swells with exudate or forms a gel.	Impermeable to exudate, microorganisms, and oxygen. Moist conditions produced promote epithelialization.	Shallow or superficial wound with minimal to moderate exudate.	Wounds with dry eschar or very light exudate.

Semipermeable foam (allevyn, lyfoam)	Soft, open cell; hydrophobic, polyurethane foam sheet 6–8 mm thick. Cells of the foam are designed to absorb liquid by capillary action.	Permeable to gases and water vapor but not to aqueous solutions and exudate. Absorbs blood and tissue fluids while the aqueous component evaporates through the dressing. Cellular debris and proteinaceous material are trapped.	Used for leg and decubitus ulcers, sutured wounds, burns, and donor sites.	Not to be applied to wounds covered with a dry scab or hard black necrotic tissue.
Amorphous hydrogel (carrasyn, royl derm, dermassist gel, hyfil, biolex, carraborb M, woundres collagen, duoderm hydroactive gel, dermagran, curasol, restore, stericare, nugel, curafil, spand-gel, intrasite, elta dermal)	Similar in composition to sheet hydrogels in their polymer and water make-up. Amorphous gels are not cross-linked. They usually contain small quantities of added ingredients such as collagen, alginate, cooper ions, peptides, and polysaccharides.	Gels clear, yellowish, or blue from copper ions. Viscosity of the gel varies with body temperature. Available as tubes, foil packets, and impregnated gauze sponges.	Used for full-thickness wounds to maintain hydration. It may be used on infected wound or as wound filler.	Heavily draining wounds. Improper use may lead to periwound maceration.

(continued)

Table 2.1. (Continued)

Dressing and fiber type	Description	Properties	Indications	Contraindications
Fillers (kaltostat, debrisan beads)	Calcium alginate that consists of an absorbent fibrous fleece with sodium and calcium salts of alginic acid (ratio 80:20). Dextranomer beads consist of are circular beads, 0.1–0.3 mm in diameter, when dry. The bead is a three-dimensional cross-linked dextran and long chain polysaccharide.	Heavily exudating wounds including chronic wounds as leg ulcers, pressure sores, fungating carcinomas. Wounds containing soft yellow slough, including infected surgical or posttraumatic wounds.	Heavily exudating wounds.	Minimally exudating wounds.
Contact layer dressings (Tulle gauze with petroleum jelly)	Greasy gauzes consisting of Tulle gauze and petroleum jelly. Dressing sheet silicone impregnated which consists of an elastic transparent polyamide net that is impregnated with a medical grade cross-linked silicone.	The dressing that is porous non-absorbent and inert is designed to allow the passage of wound exudate for absorption by a secondary dressing.	Shallow or superficial wounds with minimal to moderate exudate.	Not recommended for cleaning the wound.

| Gauze packing | Cotton gauze used both as a primary and secondary wound dressing. Gauze is manufactured as bandages, sponges, tubular bandages, and stockings. Improvement in low-linting and absorbent properties. Gauze is still a standard of care for chronic wounds. | Cotton gauze may be wetted with saline solution to confer moist properties. Possess a slight negative charge that facilitates uptake of cationic proteases. Absorbent and elastic for mobile body surfaces. | For chronic wounds it fills deep wound defects and is useful over wound gel to maintain moist wound; needs to be packed lightly. May traumatize wound if allowed to try. | Avoid multiple small dry dressing wads in wound cavity. |
| Wound vacuum assisted closure | Polyurethane foam accompanied by vacuum negative pressure in the wound bed. | Wound filled with foam and sealed with a film. Vacuum tubing is inserted and used continuously or as | Deep wound to stimulate the growth of granulation tissue. | Infected wounds and wounds with fistulae. |

synthesis and degradation, cell division, movement, and metabolism. Cytokines are regulators of inflammation and have potent stimulatory and inhibitory action on inflammatory cells. Chemokines are proteins and peptides that regulate the trafficking of leukocytes, activate inflammatory cells as neutrophils, lymphocytes, and macrophages. However, the following quote taken from a current special issue of Wound Regeneration and Repair reflects the current relationship of growth factors, cytokines, and chemokines:

Growth factors, cytokines, and chemokines are key molecular regulators of wound healing. They are all proteins, or polypeptides, and are typically synthesized and released locally, and primarily influence cells by paracrine actions. The initial concepts that growth factors were mitogens only for wound cells, that cytokines regulated inflammatory cells, and that chemokines only regulated chemoattraction of inflammatory cells were too narrow and it is now recognized that there are substantial overlaps in target cell specificity and actions between these three groups [17].

Central to understanding the future of wound dressing fibers in wound healing is an understanding of how progress in wound healing science is reshaping the design of wound dressings. Wound healing is a complex cascade of molecular and cellular events [10]. During the coagulation phase following injury, platelets initiate healing through the release of growth factors, which diffuse from the wound to recruit inflammatory cells to the wound. Thus, growth factors are responsible for the activation of immune cells, extracellular matrix deposition, collagen synthesis, and keratinocyte and fibroblast proliferation and migration. Neutrophils arrive on the scene early, and serve to clear the wound of bacteria and cellular debris. The arrival of neutrophils marks the onset of the inflammatory phase of wound healing and under acute healing conditions lasts only a few days. However, in the chronic wound the period of growing neutrophil population is extended indefinitely. Inflammation is the second phase of healing and it is mostly regulated by cytokines that are secreted by macrophages. Cytokines control cellular chemotaxis, proliferation, and differentiation. Macrophages also migrate to the wound site to destroy bacteria. However, an overabundance of cytokines and neutrophils prolong the inflammatory phase and has a negative influence on healing. Granulation tissue, which consists of fibroblasts, epithelial cells, and vascular endothelial cells, is formed about 5 days after injury. Fibroplasia is the last restorative stage of healing. Fibroplasia involves the combined effect of reepithelialization, angiogenesis, and connective tissue growth and it has been termed "a dynamic reciprocity of fibroblasts, cytokines, and extracellular matrix proteins". In a healthy person healing occurs in 21 days from coagulation, and the remodeling phase consisting of scar transformation based on collagen synthesis continues for months following injury.

When wounds fail to heal, the molecular and cellular environment of the chronic wound requires conversion to an acute wound so the ordinary sequential phases of wound healing can proceed. In June 2002, a meeting of wound healing experts formulated an overview of the current status, role, and key elements of wound bed preparation [17]. The subsequent reports in the literature from this meeting articulate well the concept of a systematic approach to wound bed preparation, which is based on an emphasis to decrease inflammatory cytokines and protease activity while increasing growth factor activity. Thus, a challenge of current wound dressing development is to promote the clinical action of wound bed preparation through addressing issues of high protease and cytokine levels and increasing growth factor levels.

2.2 The origins of moist wound dressings and the ideal wound dressing

The concept that wounds heal best when kept dry was chiefly espoused in wound management up until the late fifties because it was thought that bacterial infection could best be prevented by absorbing and removing all wound exudate. Consequently, most wounds were treated with cotton or viscose fiber material under dry conditions. However, in the early sixties Winter [6] and Hinman and Maiback [7] showed that the rate of reepithelialization increases in a moist wound versus a wound kept dry.

Occlusion is a concept in wound management that prompted a revolution during the 1970s in the production of new types of wound dressings that are still being developed. Occlusion is the regulation of water vapor and gases from a wound to the atmosphere promoting a moist environment, which allows epidermal barrier function to be rapidly restored. However, wound occlusion does require careful regulation of the moisture balance of the wound with vapor permeability to avoid exceeding the absorbency limits of the dressing. Thus, the occlusive dressing types have been developed depending on the nature of the wound and accompanying wound exudate as illustrated in Figure 2.2. The theory of moist wound healing led to approximately eight to nine separate types of wound dressing materials and devices (Table 2.1) useful for different wound treatment indications. Each of the material types that represent these distinct groups have molecular and mechanical characteristics that confer properties to promote healing under specifically defined clinical indications. For example, it has been recommended that wounds with minimal to mild exudate be dressed with hydrocolloid, polyurethane, and saline gauze, and wounds with moderate to heavy exudate be dressed with alginate dressings. Dressings may also be selected based on wound tissue color, infection, and pressure ulcer grade [11].

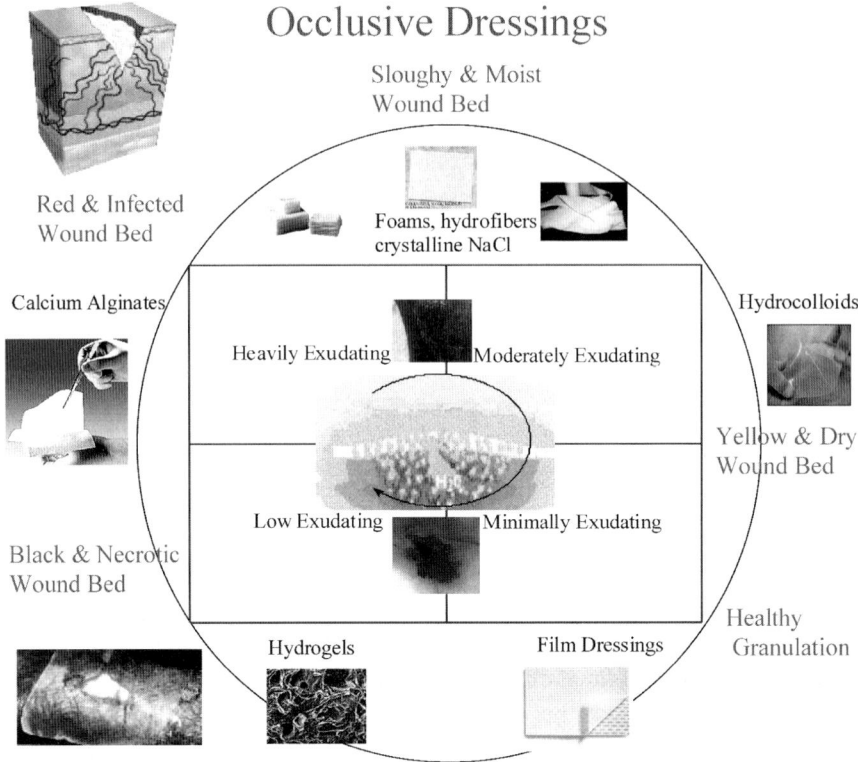

Figure 2.2. Occlusive dressings promote moist wound healing by regulating water vapor and gases in the chronic wound environment. The selection of an occlusive wound dressing depends on the degree of hydration in and around the wound tissue, its color, the presence of infection, and the pressure ulcer grade. Table 2.1 discusses the design, composition, and indications of different classes of occlusive wound dressings. For an in-depth treatment on selecting occlusive dressings see *Occlusive wound dressings. Why, when, which? By Vincent Falanga, Arch. Dermatol. 1988, June; 124(6), 872–877*

When taken as a composite of material characteristics the combined properties of the dressing materials given in Table 2.1 would approximate an ideal wound dressing. A comparison of some of the ideal properties found in both cotton and alginate wound dressings are outlined in Table 2.2. Combination of cotton and alginate in a dressing material has been reported and represents an attempt to integrate properties found in each of these two types of dressings into a single dressing [12]. Improvements in wound dressings that function at a molecular or cellular level to accelerate wound healing or monitor wound function are included among the ideal characteristics and may be termed

Table 2.2. Some ideal properties of a wound dressing as compared between cotton and alginate materials. (G) Good, (E) Excellent, and (P) Poor

Comparative properties of alginate (A) and cotton (C) dressings C A	C A		C A
Absorbency	(G) (E)	Ease of application and removal	(G/P) (E)
Adherence	(G) (E)	Elasticity	(E) (P)
Bacterial barrier	(G) (G)	Gaseous exchange	(G) (G)
Comfort	(G) (E)	Hemostatic	(G) (G)
Conformability	(G) (E)	Non-antigenic and non-toxic	(E) (E)
Drug delivery	(G) (G)	Sterilizability	(E) (E)
Durability	(E) (G)	Water vapor transmissibility	(G) (E)

interactive and intelligent wound dressings, respectively. For example, a wound dressing that removes harmful proteases from the wound to enhance cell proliferation is an example of an interactive wound dressing. A dressing having a detection device in the material signaling "time-to-change" from a defined colorimetric reaction as a molecular signal that the dressing has reached capacity of deleterious protein levels, and pH or temperature imbalance may be termed "intelligent". It seems likely that the future development of intelligent wound dressings that give beneficial clinical information on the wounds healing status will be in sync with the development of interactive dressings that perform a specific molecular or cellular function in the complex cellular and biochemical wound environment.

2.3 Interactive chronic wound dressings

The design and preparation of interactive chronic wound dressings [13] have become increasingly important as part of a solution to addressing the critical worldwide health crisis of the growing number of chronic wound patients. In the United States alone, there are over five million patients a year who suffer from chronic wounds due to the formation of decubitus bedsores brought on in the elderly nursing home or spinal chord paralysis patient. In addition, diabetes accounts for at least 60,000 patients annually who also suffer with foot ulcers. Since the mid 1990s, the number of wound care products in the well-recognized groups outlined in Table 2.1 has expanded and new groups of products have also been marketed including tissue-engineered products [14]. Recent efforts to develop wound dressings that do more than simply offer a moist wound environment for better healing have prompted most major wound dressing companies to develop research and approaches on interactive chronic wound dressings. Interactive chronic wound dressings, which possess a

Table 2.3. Carbohydrate wound dressings that stimulate growth factors and cytokines

Carbohydrate source	Associated wound dressings	Growth factor/cytokine induced and cell source	Activity	Wound healing events
Alginate [16]	Guluronic:mannuronic (80:20) dressing.	TNF-α, IL-β, IL-6 macrophages and monocytes	Collagen synthesis fibroblast and keratinocyte chemotaxis	Pro-inflammatory stimulus
DEAE sephadex [21, 22]	DEAE sephadex beads in PEG (10 mg/mL)	TGF-β platelets fibroblasts macrophages	Fibroblast activation ECM deposition, collagen synthesis TIMP synthesis MMP synthesis angiogenesis	Bead pocket increases wound breaking strength
Honey [1, 2]	Manuka and jelly bush—containing products	TNF-α, IL1β, IL-6 macrophages PMNs fibroblasts	Fibroblast and keratinocyte proliferation and chemotaxis.	Antibacterial pro-inflammatory
Aloe vera [32]	Aloeride/acemannan B—containing gels	IL-1β, TNF-α IFN-γ macrophages PMNs fibroblasts	Fibroblasts macrophages	Reduces acute radiation-induced skin reactions

mechanism-based mode of action, are targeted to biochemical events associated with pathogenesis of the chronic wound and are a part of good wound bed management.

Skin substitutes, which are being increasingly used, contain both cellular and acellular components that appear to release or stimulate important cytokines and growth factors that have been associated with accelerated wound healing [15]. Some basic materials may also play a role in up-regulating growth factor and cytokine production and blocking destructive proteolysis. In this regard, the biochemical and cellular interactions that promote more optimal wound healing have only recently been elucidated for some of the occlusive dressings described in Table 2.1. Some carbohydrate-based wound dressings that stimulate growth factor and cytokine production are outlined in Table 2.3. For example, certain types of alginate dressings have been shown to activate human macrophages to secrete pro-inflammatory cytokines associated with accelerated healing [16]. Interactive wound dressing materials may also be designed with the purpose of either entrapping or sequestering molecules from the wound bed and removing the deleterious activity from the wound bed as the wound dressing is removed, or stimulating the production of beneficial growth factors and cytokines through unique material properties. They may also be employed to improve recombinant growth factor applications. Impetus for material design of these dressings derives from advances in the understanding of the cellular and biochemical mechanisms underlying wound healing. With an improved understanding of the interaction of cytokines, growth factors, and proteases in acute and chronic wounds [17–20], the molecular modes of action have been elucidated for dressing designs as balancing the biochemical events of inflammation in the chronic wound and accelerating healing. The use of polysaccharides, collagen, and synthetic polymers in the design of new fibrous materials that optimize wound healing at the molecular level has stimulated research on dressing material interaction with wound cytokines [16], growth factors [21, 22], proteases [23, 24, 25, 29], reactive oxygen species [26], and extracellular matrix proteins [27].

2.3.1 Sequestration of wound proteases and approaches to treating chronic dermal ulcers

The prolonged inflammatory phase characteristic of chronic wounds results in an over exuberant response of neutrophils, which contain proteases and free radical generating enzymes that have been implicated in mediating much of the tissue damage associated with chronic inflammatory diseases. Since neutrophils mediate a variety of chemotactic, proteolytic, and oxidative events that have destructive activities in the chronic wound, therapeutic interventions have been proposed based on the proteolytic and oxidative mechanisms of neutrophil activity in the wound. Neutrophils contain both matrix metalloproteases and cationic serine proteases, which are two families of proteases that have been associated with a variety of inflammatory diseases, and have been implicated as destructive proteases that impede wound healing. The presence of elevated levels of these proteases in non-healing wounds has been associated with the degradation of important growth factors and fibronectin necessary for wound healing [28]. There is also a synergistic effect of further oxidative inactivation of endogenous protease inhibitors, which leads to unchecked protease activity.

A protease sequestrant dressing's design for activity may be couched in a number of molecular motifs based on the structural features of the protease, which interferes with the healing process. The molecular features of the material may be targeted to the protein's size, charge, active site, and conformation to enhance selective binding of the protein to the dressing material and removal of the detrimental protein from the wound bed. Active wound dressings that have been designed to redress the biochemical imbalance of the chronic wound in this manner are composed of collagen and oxidized regenerated cellulose [23], nanocrystalline silver-coated high-density polyethylene [29], deferrioxamine-linked cellulose [30], and electrophilic and ionically derivatized cotton [24].

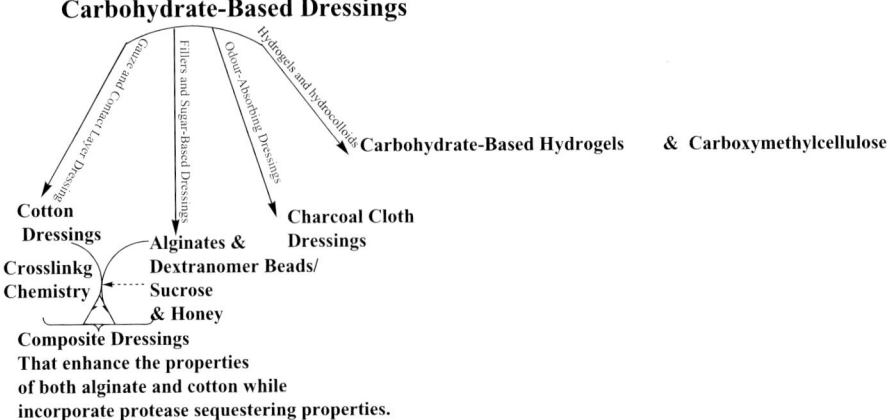

Figure 2.3. Schematic of some types of carbohydrate-based dressings and the application of cross-linking chemistry to combine two families of carbohydrate-based dressings into a more ideal composite dressing [12]

2.4 Carbohydrate-based wound dressings

Carbohydrate-based wound dressings (Figure 2.3) have received increased attention in recent years for their mechanical and molecular interactive properties with chronic and burn wounds. Traditionally, the use of carbohydrate-based wound dressings including cotton, xerogels, charcoal cloth, alginates, chitosan, and hydrogels has afforded properties such as absorbency, ease of application and removal, bacterial protection, fluid balance, occlusion, and elasticity. Focus will be given here to the design, preparation, and assessment of carbohydrate-based wound dressings as an effort to improve cotton medical textiles.

2.5 Prototype design of active cotton wound dressings

Cotton gauze has been manufactured and utilized for the last two centuries as a standard wound dressing in the care of both acute and chronic wounds. Although it is still used in much the same manner as originally conceived, there have been some fiber modifications that have improved its quality and versatility in medical applications.

The protease human neutrophil elastase found in high concentration in the chronic wound creates considerable protein destruction and prevents the wound from healing [25]. The design of wound dressings that selectively sequester proteases from the chronic wound is couched in the concept that molecular

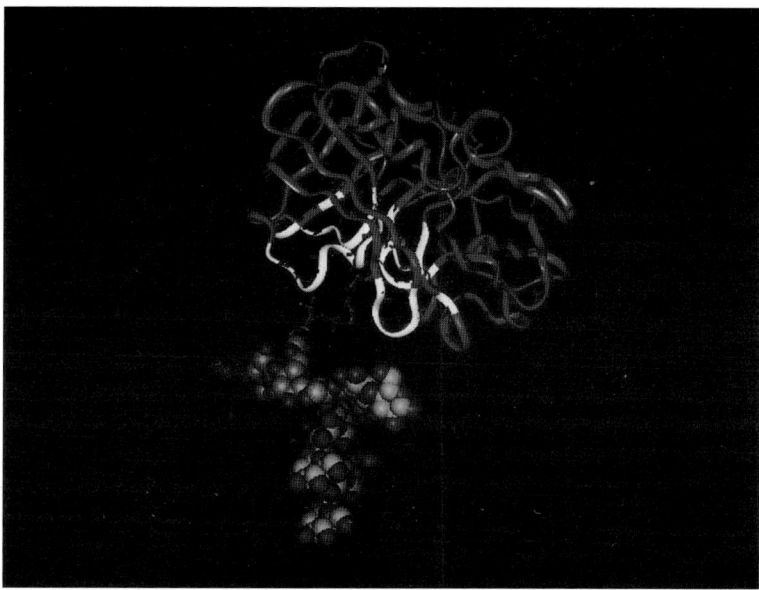

Figure 2.4. Computer graphic model of peptide-bound cellulose docked at the active site of human neutrophil elastase. Cellulose is depicted as the green and red CPK model. The peptide portion of the conjugate is a ball and stick model shown docked within the yellow highlighted ribbon depicting the active site of elastase

features and properties of the protease can be used to tailor the molecular design of the cotton fiber needed for selective sequestration of the protease. Thus, the enzyme size, overall charge, and active site mechanism for binding substrate may be employed to create the appropriate fiber design that might best bind the enzyme selectively. The design approach of the prototype for selective sequestration is a molecular model of a cellulose conjugate containing an active site recognition sequence docked to the active site of human neutrophil elastase as shown in Figure 2.4. The subsites of enzyme active site interaction consist of the sequence conjugate H-Val-Pro-Glycine-O-ester-Cellulose.

2.6 Preparation and assay of the prototype active cotton-based wound dressing

The preparation of the prototype cotton wound dressing containing the conjugate shown in Figure 2.4 required synthesis of a tripeptide sequence on the cotton fiber. This peptide sequence was linked to the cellulose of the cotton fiber at both ends of the peptide sequence and tested for activity to sequester human neutrophil elastase [25]. Assay of the peptide conjugate on cotton was

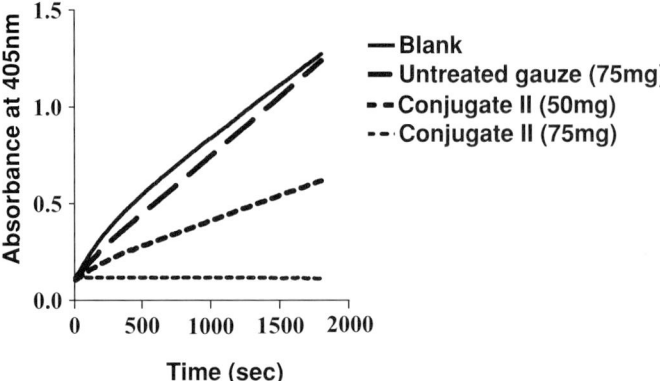

Figure 2.5. Reaction progress curves for a peptide conjugate [25] as illustrated in the molecular model in Figure 2.2 in solutions of elastase that have been treated with interactive cotton wound dressing fibers. The reaction is an enzyme hydrolysis of the elastase substrate left in solution following treatment. A slower reaction as seen with the peptido–cellulose conjugates versus the untreated cotton dressing reflects increased removal of elastase from solution

completed by incubating the cotton wound dressing in a solution of elastase for an hour and assessing the sequestration activity of the modified wound dressing fiber. Determination of the amount of enzyme taken up by the fiber was based on the kinetic profile of the reaction progress curve of enzyme remaining in solution and its reaction with substrate as shown in Figure 2.5. The elastase substrate is employed in the assay as a putative protein associated with healing. Thus, a smaller amount of substrate left in solution and a slower reaction progress curve is associated with higher levels of activity bound to the dressing and a more active wound-dressing fiber. It is noteworthy in this regard that the activity of the peptide conjugate fiber is dose dependent.

2.7 Design of active cotton-based wound dressings

The design of the chronic wound dressing requires a simple, economically feasible modification that is imparted to the cotton fiber in a one- or two-step aqueous finishing technique. This is necessary so economical methods can be adopted in the textile mill to modify the cotton. An active site protease sequestrant would be based on the potential for the modified fiber to interact analogous to an enzyme inhibitor or substrate as shown in the previous peptide conjugate of cellulose example, and as illustrated in Figure 2.4. On the other hand, a charge sequestrant material is based on binding of the enzyme to the cotton fiber through ion pairing: elastase is positively charged, thus a negatively

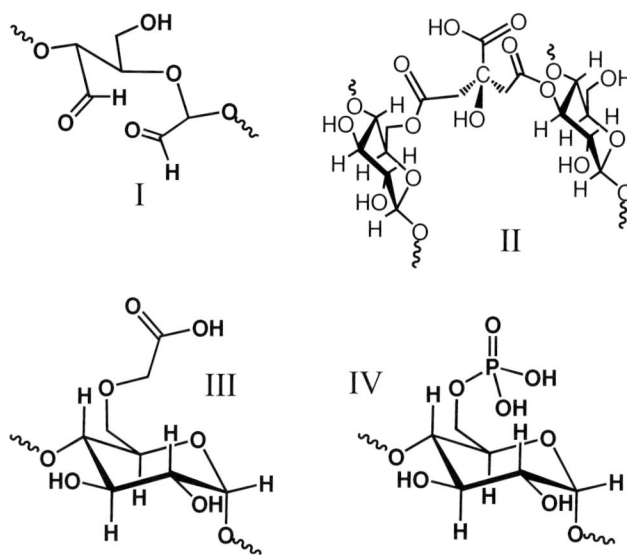

Figure 2.6. Representative structures of modified anhydroglucose monomer units in the cellulose chain upon treatment of the cotton gauze. I, structure of dialdehyde cotton cellulose; II, poly-carboxylic acid cross-linked cotton; III, carboxymethylcellulose cotton; and IV, phosphorylated cotton

charged fiber would ion pair with the enzyme. The types of sequestrant motifs might be based on structures of modified cellulose as shown in Figure 2.6. These cotton cellulose modifications are dialdehyde, carboxymethylated, phosphorylated, and polycarboxylate cross-linked modifications. The active site sequestrant is the dialdehyde functional group (I), and the negatively charged modifications are the two forms of carboxylated and phosphorylated cellulose (II–IV). The preparation of these functionally finished cotton wound dressings has been previously reported [24].

The proposed mechanism of action of the dialdehyde cotton wound dressing is shown in Figure 2.7. The proposed mechanism for sequestration is thought to occur by formation of a hemiacetal through attack of the Ser-195 within the active site of residue with assistance from Histidine-57 and Aspartate-102. The concerted interaction of these residues termed the catalytic triad of the serine protease leads to cleavage of a peptide bond when proteolytic activity occurs, but in this model the interaction is more similar to inhibitor binding of the enzyme. To show that the dialdehyde cotton may function to sequester the elastase *via* this molecular mechanism the enzyme has been assayed with a soluble form of dialdehyde starch which best approximates properties of cotton as a carbohydrate in solution.

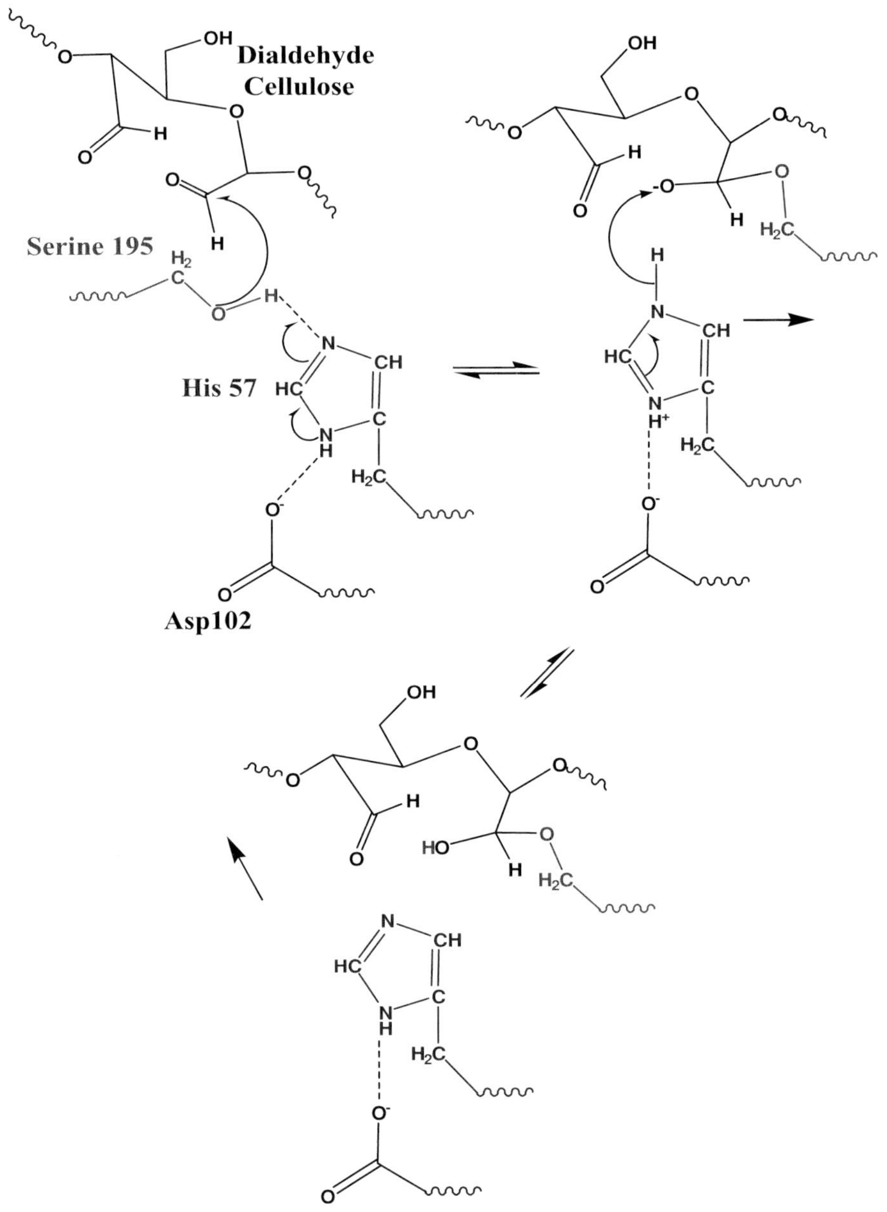

Figure 2.7. The postulated mechanism of action for protease sequestration by dialdehyde cotton [24] as an example of a modified fiber sequestrant acting at the active site of the protease to enable enhanced protease removal from the chronic wound. The broad range proteolytic activity of serine proteases released by neutrophils into the wound environment is responsible in part for the degradation of growth factors and extracellular matrix proteins. The active site of catalytic activity in serine proteases consists of a catalytic triad shown by X-ray studies to consist of a charge relay among amino acids for substrate hydrolysis. Inhibition of the active site occurs

2.8 Understanding and predicting how active cotton wound dressings may perform in the chronic wound

There is some controversy concerning the usefulness of animal models in testing chronic wound dressings for efficacy. The pathology of the chronic wound is even more complex than the healing wound, and difficult to mimic. One predictor for efficacy of a chronic wound dressing is testing the dressing *in vitro* with chronic wound fluid or proteins that mimic the environment and protein concentration as well as makeup of chronic wound exudate. During the course of developing a modified cotton product for commercialization, two models for studying the performance of the modified cotton fiber under conditions that mimic chronic wound fluid exudate were made. One model consisted of assaying the modified fiber in diluted chronic wound fluid containing high elastase activity similar to that of the chronic wound [24]. More recently, we have developed a model utilizing albumin concentrations that mimic those levels of albumin found in the chronic wound in the presence of elastase. Another purpose in utilizing the albumin model is to better understand how albumin may compete for binding sites on different functional groups of modified cotton cellulose, and compare capacities for competitive protease binding. Using these types of models, we have begun to study and compare more closely the mechanisms for competitive binding through ion pairing between the enzyme and cotton as shown in Figure 2.5 with the "inhibitor-active site". Figure 2.8 shows the results of an experiment designed to evaluate the capacity of a type of charge sequestrant wound dressing currently in development that removes elastase from solution. The results of this 24-hour assay where the dressing is challenged with a constant concentration of elastase and evaluated for continued removal of the protease from solution suggest good capacity of the charge sequestrant wound-dressing motif.

2.9 Summary

Wound care products along with the clinical practice of wound care itself have rapidly matured over the past 20 years and become a molecular-biotechnology focused industry. The product market is now valued at $1.74 billion, and five million Americans suffering from chronic open wounds

---

Figure 2.7. (Continued) analogous to substrate peptide bond cleavage when proton transfer from Serine-195 is transferred to Histidine-57 upon attack of the Serine-195 hydroxyl oxygen at the electrophilic carbonyl of the anydrogluco-aldehye. Several classes of aldehyde and ketone-based inhibitors have been developed (*Edwards, P. D.; Bernstein, P. R. Synthetic inhibitors of elastase. Med. Res. Rev. **1994**, 14, 127–194*) for a variety of inflammatory diseases but few have been adapted to wound dressing

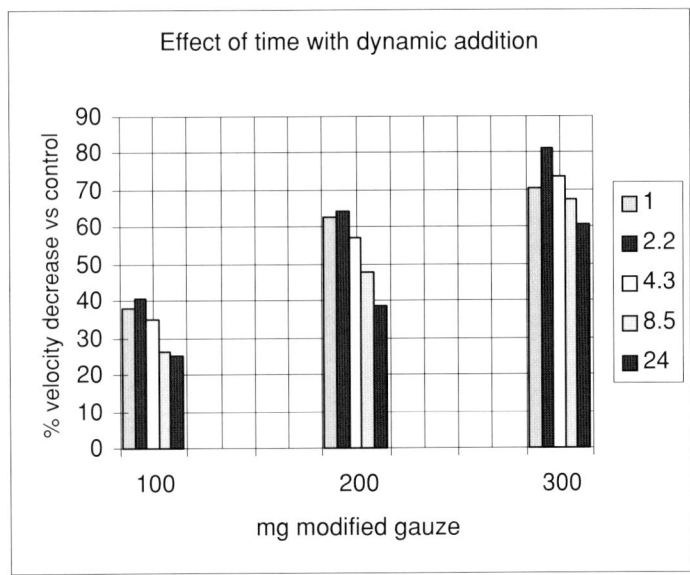

Figure 2.8. Results of a dynamic addition capacity study for a charge sequestrant dressing. The bar graph is a plot of percent decrease in elastase activity versus varying amounts of modified gauze (rep100 mg, 200 mg, and 300 mg) used over a 24-h period. Elastase levels are regenerated through out the 24-h time course to challenge the dressing material with increasing levels. A solution mimicking wound fluid was prepared consisting of 4% albumin and 2 milliunits of elastase per milligram of protein. The results show that the dressing continues to remove elastase after 24 h in the presence of protein levels found in the chronic wound

require care that is estimated at $5–7 billion per year and increasing at an annual rate of 10% [31]. Research and development is currently underway to achieve more ideal wound dressings. As shown in the cartoon in Figure 2.9, the chronic wound dressing of the future will probably have structural features built into a single dressing. This prototype dressing would confer properties of moisture balance, protease sequestration, growth factor stimulation, "time-to-change indicator", antimicrobial activity, and oxygen permeability. Many of these properties are already present in current wound dressings; however, no single wound dressing product offers all of them. The future success of wound care products from modified traditional materials or new materials depends on continued mechanism-based research at all levels from basic through clinical assessment. As new products like those included in the interactive wound dressing category continue to become available evidence regarding their relative efficacy will be needed to provide the wound care practitioner with data in making the best product selection for the patients needs.

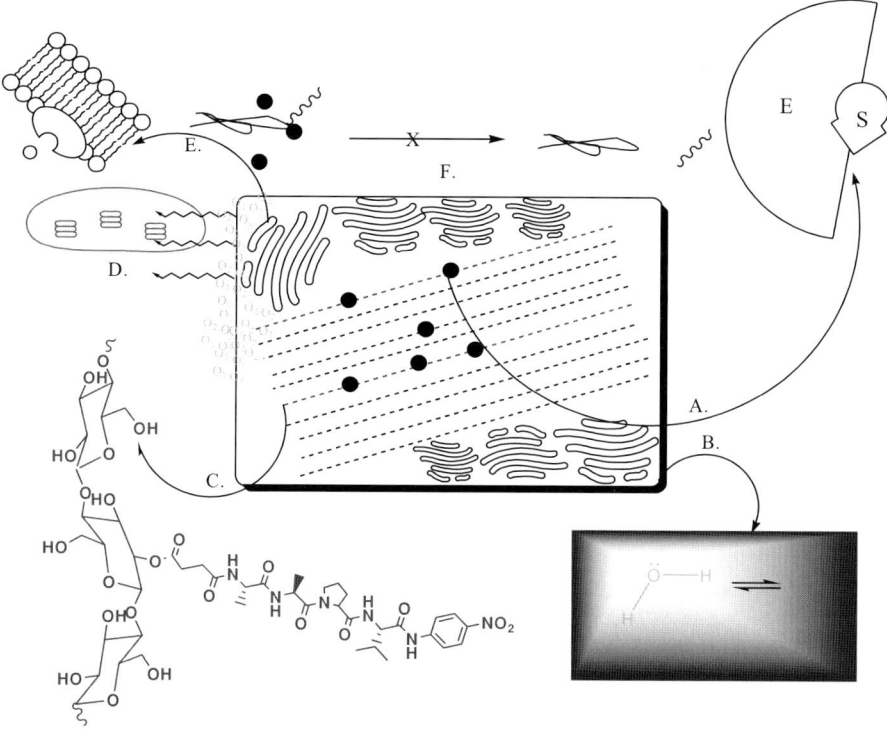

Figure 2.9. Cartoon of some of the structural properties of a wound dressing of the future. (**A**) It reduces proteases and their activity toward the degradation of growth factors (F.) by selectively binding proteases similar to the tailored fit of an enzyme–substrate complex [ES]; (**B**) It possesses absorbency that responds to exudate of the wound environment by adjusting the moist wound environment in equilibrium with optimal wound moisture for healing; (**C**) A colorimetric indicator signals the wound dressing has reached its capacity to redress biochemical imbalance in the chronic wound. A peptide-containing chromophore built into the dressing fiber might release a colorimetric signal in response to reaching its capacity to perform as a protease sequestrant. (**D**) The dressing possesses antimicrobial activity and serves as a barrier to wound contamination while remaining permeable to oxygen. (**E**) and (**F**) The dressing optimally stimulates the production of growth factors and cytokines in the wound environment while preventing their degradation. Consequently, growth factors trigger their membrane bound receptors, and proteases are blocked from degrading growth factors and their cellular receptors

References

1. Cooper, R. A.; Molan, P. C.; Harding, K. G. Antibacterial activity of honey against strains of *Staphylococcus aureus* from infected wounds. *J. R. Soc. Med.* **1999**, *92*, 283–285.
2. Tonks, A. J.; Cooper, R. A.; Jones, K. P.; Blair, S.; Parton, J.; Tonks, A. Honey stimulated inflammatory cytokine production from monocytes. *Cytokine* **2003**, *21*, 242–247.
3. Thang, P. T.; Patrick, S.; Teik, L. S.; Yung, C. S. Anti-oxidant effects of the extracts from the leaves of *Chromolaena odorata* on human dermal fibroblasts and epidermal keratinocytes

against hydrogen peroxide and hypoxanthine-xanthine oxidase induce damage. *Burns* **2001**, *27*, 319–327.

4. Wool offers burns breakthrough, in 'Beyond the Bale', Issue 11, May 2004.

5. Elliot, I. M. Z. *A Short History of Surgical Dressings*. Pharmaceutical Press, London, **1964**.

6. Winter, G. D. Formation of the scab and the rate of epithelization of superficial wounds in the skin of the young domestic pig. *Nature* **1962**, *193*, 294.

7. Hinman, C. D.; Maibach, H. Effect of air exposure and occlusion on experimental human skin wounds. *Nature* **1963**, *200*, 377.

8. Rheinwald, J. G.; Green, H. Formation of a keratinizing epithelium in culture by a cloned cell line teratoma. *Cell* **1975**, *6*, 317–330.

9. Enquist, I. F. The principles of wound healing. In: Davis, L. (Ed.) *Christopher Textbook of Surgery*. **1968**.

10. Clark, R. A. F. Wound repair: Overview and general considerations. In: Richard, A. F. C. (Ed.) *The Molecular and Cellular Biology of Wound Repair*. Plenum Press, New York, **1996**, pp. 3–35.

11. Bello Y. M.; Phillips, T. J. Recent advance in wound healing. *JAMA* **2000**, *283*(6), 716–718.

12. Edwards, J. V.; Bopp, A. F.; Batiste, S. L.; Goynes, W. R. Human neutrophil elastase inhibition with a novel cotton-alginate wound dressing formulation. *J. Biomed. Mater. Res.* **2003**, 433–440.

13. Draft Guidance for the Preparation of an IDE Submission for an Interactive Wound and Burn Dressing, in 817 Guidance for Interactive Wound Dressings, U.S. Food and Drug Administration Center for Devices and Radiological Health, p1–5, 1995.

14. Morgan, D. Wounds—what should a dressing formulary include. *Hosp. Pharm.* **2002**, *9*, 261–266.

15. Falling, V.; Isaac's, C.; Packet, D.; Downing, G.; Kowtow, N.; Butter, E. B.; Harden-Young, J. Wounding of bioengineer skin: Cellular and molecular aspects after injury. *J. Invest. Dermatol.* **2002**, *119*, 653–660.

16. Thomas, A.; Harding, K. G.; Moore, K. Alginates from wound dressings activate human macrophages to secrete tumor necrosis factor-α. *Biomaterials* **2000**, *21*, 1797–1802.

17. Schultz, G. S.; Sibbald, R. G.; Falanga, V.; Ayello, E. A.; Dowsett, C.; Harding, K.; Romanelli, M.; Stacey, MC.; Teot, L.; Vanscheidt, W. Wound bed preparation: A systematic approach to wound management. *Wound Repair Regen.* **2003**, *11*, 1–28.

18. Mast, B. A.; Schultz, G. S. Interactions of cytokines, growth factors and proteases in acute and chronic wounds. *Wound Repair Regen.* **1996**, *4*, 411–420.

19. Baker, E. A.; Leaper, D. J. Proteinases, their inhibitors, and cytokine profiles in acute wound fluid. *Wound Repair Regen.* **2000**, *8*, 392–398.

20. Trengove, N. J.; Bielefeldt-Ohmann, H.; Stacey, M. C. Mitogenic activity and cytokine levels in non-healing and healing chronic leg ulcers.

21. Christoforou, C.; Lin, X.; Bennett, S.; Connors, D.; Skalla, W.; Mustoe T.; Linehan, J.; Arnold, F.; Guskin, E. Biodegradable positively charged ion exchange beads: A novel biomaterial for enhancing soft tissue repair. *J. Biomed. Mater. Res.* **1998**, *42*, 376–386.

22. Connors, D.; Gies, D.; Lin, H.; Gruskin E.; Mustoe, T. A.; Tawil, N. J. Increase in wound breaking strength in rats in the presence of positively charged dextran beads correlates with an increase in endogenous transforming growth factor-β 1 and its receptor TGF βR1 in close proximity to the wound. *Wound Repair Regen.* **2000**, *8*, 292–303.

23. Cullen, B.; Smith, R.; McCulloch, E.; Silcock, D.; Morrison, L. Mechanism of action of PROMOGRAN, a protease modulating matrix, for the treatment of diabetic foot ulcers. *Wound Repair Regen.* **2002**, *10*, 16–25.

24. Edwards, J. V.; Yager, D. R.; Cohen, I. K.; Diegelmann, R. F.; Montante, S.; Bertoniere, N.; Bopp, A. F. Modified cotton gauze dressings that selectively absorb neutrophil elastase activity in solution. *Wound Repair Regen.* **2001**, *9*, 50–58.

25. Edwards, J. V.; Batiste, S. L.; Gibbins, E. M.; Goheen, S. C. Synthesis and activity of NH2- and COOH-terminal elastase recognition sequences on cotton. *J. Peptide Res.* **1999**, *54*, 536–543.

26. Moseley, R.; Leaver, M.; Walker, M.; Waddington, R. J.; Parsons, D.; Chen, W. Y. I.; Embery, G.pt Comparison of the antioxidant properties of HYAFF- 11p75, AQUACEL and hyaluronan toward reactive oxygen species in vitro. *Biomaterials* **2002**, *23*, 2255–2264.

27. Kirker, K. R.; Luo, Y.; Nielson, J. H.; Shelby, J.; Prestwich, G. D. Glycosaminoglycan hydrogel films as bio-interactive dressings for wound healing. *Biomaterials* **2002**, *23*(17), 3661–3671.

28. Yager, D.; Nwomeh, B. The proteolytic environment of chronic wounds. *Wound Repair Regen.* **1999**, *7*, 433–441.

29. Wright, J. B.; Lam, K.; Buret, A. G.; Olson, M. E.; Burrell, R. E. Early healing events in a procine model of contaminated wounds: Effects of nanocrystalline silver on matrix metalloproteinases, cell apoptosis, and healing. *Wound Repair Regen.* **2002**, *10*, 141–151.

30. Meyer-Ingold, W.; Eichner, W.; Ettner, N.; Schink, M. Wound coverings for removal of interfering factors from wound fluid. United States Patent, 6,156,334, 2000.

31. http://www.aawm.org/news_trend.html.

32. Byeon, S. W.; Pelley, R. P.; Ullrich, S. E.; Waller, T. A.; Bucana, C. D.; Strickland, F. M. *Aloe barbadensis* extracts reduce the production of interleukin- 10 after exposure to ultraviolet radiation. *J. Invest. Dermatol.* **1998**, *110*, 811.

Chapter 3

BEHAVIOR OF CELLS CULTURED ON CUPROPHAN

N. Faucheux, J. L. Duval, J. Gekas, M. Dufresne, R. Warocquier, and M. D. Nagel

Domaine Biomatériaux-Biocompatibilité, UMR CNRS 6600, Université de Technologie de Compiègne, BP 20529, F-60205 Compiègne Cédex, France

3.1 Introduction

Controlling cell shape induced by cell–substrata interaction appears of prime importance to influence subsequent biological processes such as cell migration, proliferation, or apoptosis [1–3].

Cell shape and cytoskeletal organization may change through the transmission of mechanical stresses mediated by cell-surface integrins. Integrins are receptors composed of α and β subunits linked in a transmembrane heterodimer. Some of these mediate the adhesion of cells to Arg-Gly-Asp (RGD)-containing proteins such as vitronectin (VN) or fibronectin (FN), both of which play a major role in the attachment of cells plated out in the presence of fetal bovine serum (FBS). Adsorbed proteins maintaining their sites in conformational active structures will be recognized then by specific integrins. In this respect, cell culture treated polystyrene dishes (PS) which adsorb adhesive serum proteins without altering them favor cell spread and proliferation.

Since the surface of the material may be considered as a matrix of chemical groups influencing cell-material contact, every monomer unit of the surface is a potential site for interaction with proteins and cells. Therefore, depending on their surface properties, biomaterials will adsorb more or less adhesive proteins and initiate different cell behaviors.

J. V. Edwards et al. (eds.), Modified Fibers with Medical and Specialty Applications, 35–47.

3.2 An immediate cell-material relationship model: Cuprophan–Swiss 3T3 murine fibroblast cell line

3.2.1 A cellulose substratum: Cuprophan

3.2.1.1 Nature

Cuprophan (CU), a regenerated cellulose membrane produced by the cuprammonium process, appears to be very hydrophilic (due to OH groups in the cellobiose units of the cellulose molecule) as attested by a 22° water contact angle.

We compared its ability to adsorb serum proteins of the culture medium to the moderately wettable acrylonitrile AN69 membrane (40° contact angle). Those biomaterials are both currently used in bioreactors and hemodialyzers.

3.2.1.2 In vitro serum protein adsorption

Low ability of CU to adsorb serum adhesive proteins
Immunoblot analysis of FN and VN adsorbed on 45 min FBS-conditioned AN69 and CU was performed. FN was not detected in the eluate of CU but was resolved into two bands of 230 and 220 kDa in the eluate of AN69. VN was resolved into a faint band of 74 kDa in the eluate of CU and a very large VN band was detected in the sample from AN69. CU adsorbed 60-fold less serum proteins than AN69 in our experimental system [4].

Low binding strength of adsorbed total serum proteins to CU
SDS PAGE protein profiles of supernatants (250, 230, 170, 130, 110, 75, 66, 58, and 52 kDa) obtained from washes of 45 min FBS-conditioned AN69 and CU were analyzed. Proteins were detected in all washes of FBS-conditioned AN69 with 10%, 30%, 50%, and 70% isopropanol. In contrast, FBS proteins were completely removed from FBS-conditioned CU by the 10% and 30% isopropanol washes, attesting a weak strength of serum adsorbed proteins on CU [4].

3.3 Swiss 3T3 murine fibroblast early reaction to CU

Swiss 3T3 fibroblasts are widely used for cytocompatibility studies. We studied their behavior after 45, 90, or 180 min in culture medium with or without serum, when seeded on CU compared to AN69 membranes.

3.3.1 Cell shape observations

Cell spreading is mediated by both cell-surface integrin receptors and the small GTP-binding protein RhoA [5]. The engagement of integrins and their subsequent clustering in focal adhesion complexes lead to the generation of intracellular macromolecular binding complexes including the cytoskeletal proteins vinculin and talin. RhoA has been implicated in the regulation of integrin activation by promoting avidity modulation, a process known as inside-out signaling [6–9].

The role of serum adhesive protein adsorption was clearly evidenced by alteration of cell morphology in accordance with experimental conditions.

Forty-five minutes after seeded on CU, as well cultured in 10% FBS (low protein adsorption) as in serum-free medium (no protein adsorption), cells were rounded and aggregated. Each aggregate on CU appeared to be attached to the substratum by a few cells at the end of the aggregate. By contrast, as well on AN69 as polystyrene culture dishes, cells were flattened in the presence of serum (high protein adsorption) but rounded in serum-free medium (no protein adsorption) [10] (Figure 3.1).

Cells attached to CU and PS were fixed and stained with anti-vinculin and anti-integrin αv to assess the formation of focal adhesion complexes and with

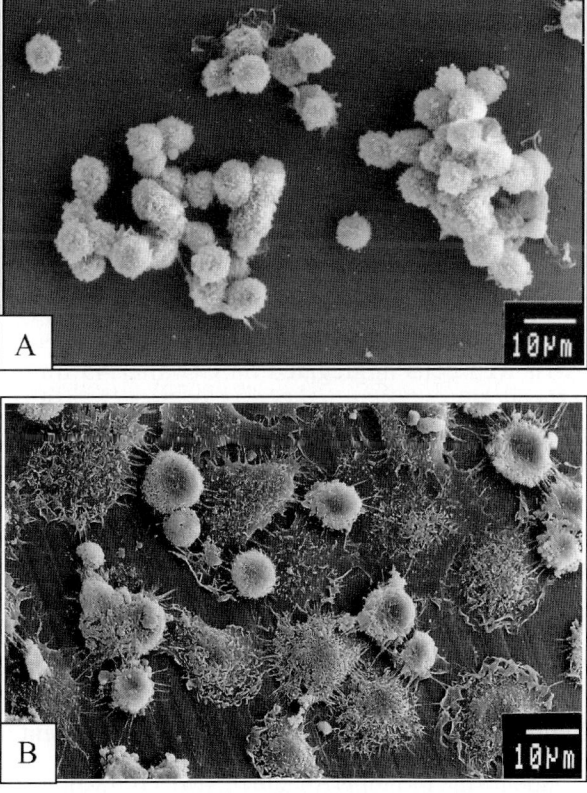

Figure 3.1. Swiss 3T3 cell morphology on CU (A) and PS (B) in medium supplemented with 10% FBS after 45 min

rhodamine-phalloidin to label the actin stress fibers after incubation for 45 and 180 min. Most of the cells attached to PS after 45 min incubation have begun to spread. There were punctuated concentrations of integrin αv and/or vinculin at the perimeters of structures that resembled focal adhesion complexes. By contrast, neither clusters of integrin αv nor punctuate concentrations of vinculin were detected in aggregated cells on CU. There were typical pattern of integrin αv rich clusters or vinculin in focal adhesion complexes in cells incubated for 180 min and attached to PS, whereas aggregated cells on CU contained few integrin αv or vinculin-positive focal adhesion complexes at the cell borders. There were some actin filaments in cells attached to the PS but no organization of the actin cytoskeleton in aggregated cells on CU after 45 min incubation. Fibroblasts attached to PS after 180 min contained many bundles of actin filaments. Two types of stress fibers were formed. Most of them were parallel, while the rest contained stellar fibers radiating from several foci. By contrast, cells aggregated on CU remained rounded and showed heavy staining circling the cell periphery with no distinct stress fibers [11].

3.3.2 Connexin 43 organization and gap junction communication

Gap junctions play a major role in the regulation of differentiation, cell growth, and the regulation of numerous metabolic processes [12–15]. Gap junctions consist of aggregated channels that directly link the interior of neighboring cells. Each gap junction channel is made up of two hemichannels or connexons, which are formed by the oligomerization of protein subunits known as connexins (Cx). The newly synthesized proteins oligomerize into connexons and are transported to the plasma membrane where they interact with connexons in adjacent cells to form cell-to-cell channels [16, 17].

Immunostaining and transmission electron microscopy observations showed that Swiss 3T3 cells aggregated on CU after 45 min incubation established short linear gap junctions composed of Cx 43 in cell-surface plaques [18] (Figure 3.2).

We then used the fluorescent tracer Lucifer Yellow (LY) to examine the ability of cells aggregated on CU to communicate *via* gap junctions. Cells seeded on CU were incubated for 90 min at 37 °C and LY was microinjected into one cell of an aggregate for 1 min. We observed a progressive diffusion of the fluorescent dye toward the neighboring cells reflecting the functionality of gap junctions [19] (Figure 3.3).

3.3.3 Early biochemical events induced by cell attachment on CU

Cyclic $3'$–$5'$-adenosine monophosphate (cAMP) is involved in a wide range of cell functions, including cell contraction and motility, but the cAMP pathway

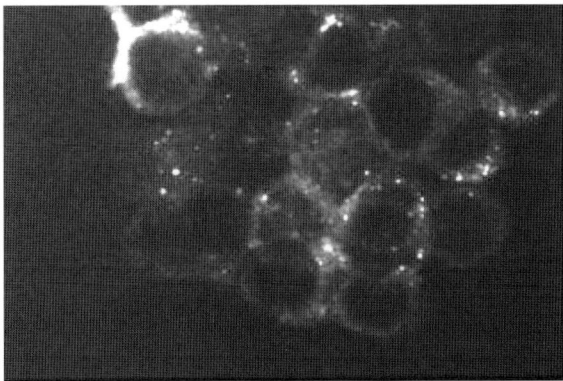

Figure 3.2. Immunostaining for connexins 43 channel clustering in Swiss 3T3 cells aggregated on CU after 45 min. Brightly fluorescent dots are detected at cell–cell apposition points. (magnification ×1250)

is especially important in cell shape, adhesion, cytoskeletal structure, and focal contact formation [20, 21]. An increase in intracellular cAMP induced for example by forskolin, a direct activator of adenylyl cyclase, the effector enzyme which catalyses the production of cAMP from adenosine triphosphate (ATP), causes marked morphological changes. It can also be a negative modulator of RhoA through the activation of protein kinase cAMP dependent (PKA) [22, 23].

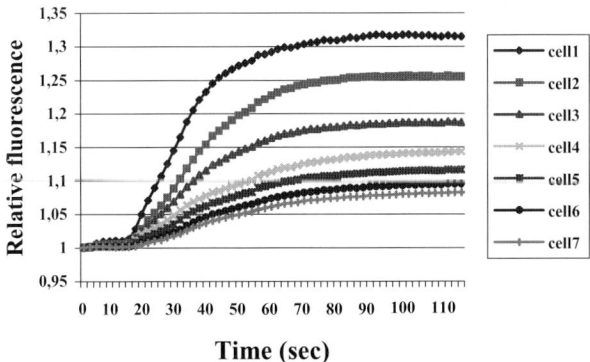

Figure 3.3. Quantification of the diffusion of LY dye in cells aggregated on CU after 90 min incubation. The fluorescent dye was injected into one cell of an aggregate. The diffusion of dye from the impaled cell into neighboring cells was viewed with epifluorescence every 2 s. An inverse relationship was observed between the increase in fluorescence intensity and the distance from injected cell. The diffusion of the dye within the neighboring cells was progressively delayed with the increase of the distance from the injected cell

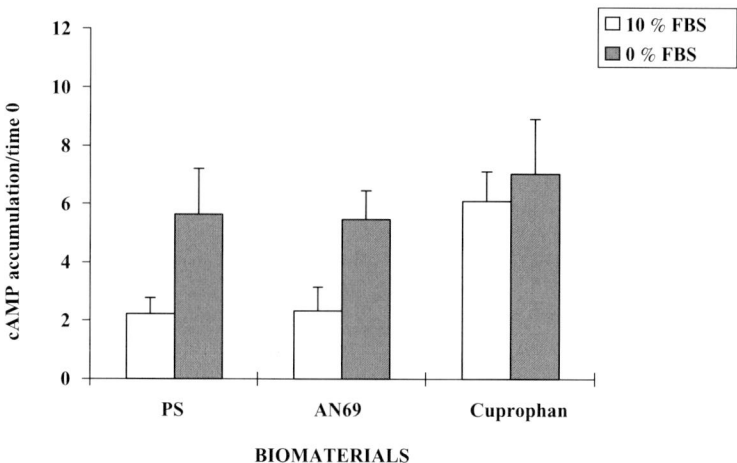

Figure 3.4. cAMP accumulation in Swiss 3T3 fibroblasts attached to CU, AN69 or PS, 45 min after seeding in medium containing 10% FBS or in serum-free medium. Results (cAMP 10^{-15} M/10^6) cells are referred to the amount of cAMP in the cells in suspension at time 0. Results are means $\pm$ SD of three experiments, each performed in triplicate

We observed, by RIA measurement, that Swiss 3T3 cells adhering for 45 min to AN69 in serum-free conditions increase their content of cAMP and become aggregated. These morphological and biochemical changes are prevented by adding 10% FBS to the medium while cells aggregated on CU remain unaffected (Figure 3.4). Furthermore, we demonstrated that the catalytic activity of adenylyl cyclase is activated by the attachment of cells to CU and that MDL 12330 A (*cis-N*-(2-phenylcyclopentyl)azacyclotridec-1-en-2-amine), a specific inhibitor of adenylyl cyclase, prevents cell aggregation and abolishes Cx 43 channel clustering on CU.

By contrast, forskolin and 8-Br-AMP (a cell-permeable analog of cAMP) causes Cx 43 clustering in cells attached to PS. Hence, Cx 43 channel clustering is regulated by cAMP in Swiss 3T3 cells [18, 23]. The surface integrins of cells on AN69 or PS bind VN and/or FN, favoring the attachment and spreading of cells under our experimental conditions. Cells on AN69 or PS have well-defined focal adhesion complexes and stress fibers [18]. This reflects the effective "outside–inside" transmembrane signaling produced by attachment of integrins to substrate-adsorbed proteins. In contrast, CU adsorbs VN poorly and does not support cell spreading or the formation of focal contacts and stress fibers. Competitively blocking VN and FN receptors with the disintegrin echistatin (a potent inhibitor of the binding of all the RGD-dependent integrins to their natural ligands) does not modify the aggregation of cells when seeding on CU.

We investigated the shift between integrin-signaling RhoA and the cAMP pathway: after incubation of 45 min, we measured the immunoreactivity of RhoA in cell membrane preparations from cells attached to CU and PS. We

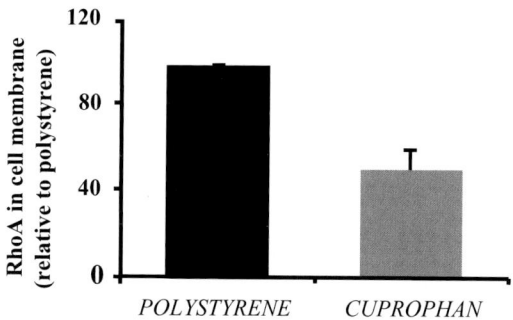

Figure 3.5. Immunoreactivity of RhoA in cell membrane preparations. Membrane proteins (75 μg) were resolved by SDS PAGE and immunoblotted with monoclonal antibody directed against RhoA. Blots were quantified by densitometry after three independent experiments

found significantly lower concentrations of RhoA in the cell membranes of cells attached to CU than in cells on PS (Figure 3.5). The effect of the cAMP pathway on RhoA was assessed using PKI, a specific inhibitor of PKA. Adding PKI to the medium increased RhoA in the plasma membrane of cells aggregated on CU [11]. The whole of the observations reported above were carried out from cells cultured on CU for 45, 90, or 180 min. Early biochemical events lead to program cell functions that are presented below.

3.4 Swiss 3T3 murine fibroblast functions on CU (cell cultures)

We focused on three well-characterized aspects of cell behavior of Swiss 3T3 cells aggregated on CU for 24 and 48 h, compared to cells cultured on PS: proliferation, protein synthesis, and programmed cell death (apoptosis).

3.4.1 Cell proliferation

Proliferation of cells cultured for 24, 48, and 72 h on CU compared to PS was assessed by counting cells in a Malassez hemocytometer (Figure 3.6) and PKH 26 staining detected by flow cytometry. Compared to cells spread on PS, proliferation of cells aggregated on CU for 24 and 48 h was significantly decreased and percentages of dead cells (stained by Trypan Blue) in the total cell population increased (4.7% $\pm$ 1.1% on PS *versus* 16.0% $\pm$ 1.9% on CU after 48 h).

3.4.2 Total protein cell content

Cells cultured for 48 h on CU and PS were removed by trypsin-EDTA and counted. Proteins were determined by the Bradford method. Cells cultured on CU contain twofold more proteins than cells on PS at 48 h (Figure 3.7).

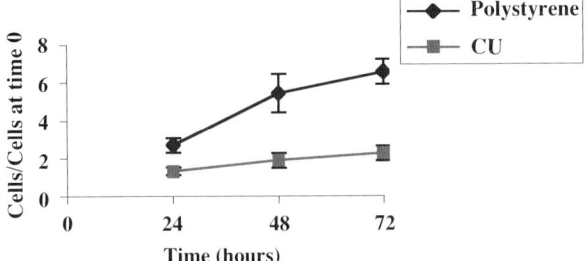

Figure 3.6. Growth of Swiss 3T3 cells cultured on CU and on PS. The growth rate differences at each timepoint were statistically significant (Kruskal–Wallis non-parametric test $p < 0.01$)

3.4.3 Apoptosis

The translocation of phosphatidylserine from the inner part to the outer layer of the plasma membrane is an early event of apoptosis [24–26]. Annexin V binding can be detected by flow cytometry to evaluate this phenomenon.

Propidium iodide negative–Annexin V positive apoptotic cells were thus quantified from cells cultured on CU and PS for 24 and 48 h. The percentage of apoptotic cells was significantly higher on CU than on PS (29.1% ± 2.7% on CU *versus* 9.3% ± 0.7% on PS after 48 h).

The initiation and progression of apoptosis caused by various stimuli depend on a family of intracellular cysteine proteases, the caspases (cysteinyl-aspartate-specific proteinases) [27–29]. We also assayed caspase 3 activity mediating the final stage of apoptosis.

Results gained with Annexin V were confirmed by colorimetric measurements of caspase 3 activity in cells cultured on CU and PS for 24 h. Caspase 3 activity was significantly increased at 24 and 48 h (Figure 3.8) and the existence of a mitochondria-dependent apoptotic process (caspase 9 activation) was demonstrated in cells aggregated on CU [30].

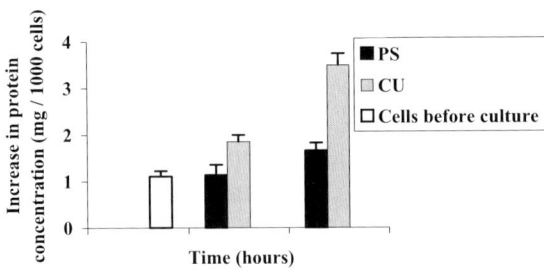

Figure 3.7. Protein quantification in total cell lysates by a colorimetric assay. The absorbance was measured at 562 nm. A significantly higher concentration of proteins was observed on CU compared to PS after incubation for 24 and 48 h (Kruskal–Wallis non-parametric test $p < 0.01$)

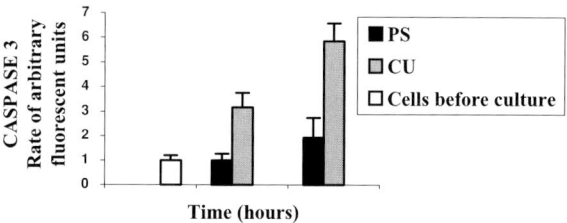

Figure 3.8. Caspase 3 activity in Swiss 3T3 fibroblasts cultured on PS or CU for 24 and 48 h. Cells cultured on CU had significantly greater caspase 3 activity than cells cultured on PS (Kruskal–Wallis non-parametric test $p < 0.001$). The activity in cells on PS was not significantly different from the basal level measured immediately upon cell seeding (time 0)

3.5 Long-term behavior of murine melanoma B16 F10 explants on CU

B16 F10 murine melanoma cells possess a highly metastatic potency when injected to mice. Tumor explants ($1–2 \, mm^3$) layered on semi-solid culture medium were covered with $0.7 \, cm^2$ CU or AN69 pieces and cultured for 14 days. Proliferation, adhesion, and migration of sorting cells were assessed from the cell layer developed around each explant. Surface area and cell counts after trypsin-EDTA detachment kinetics can give thus a good evaluation of cell behavior [31].

Proliferation and migration of melanoma cells were significantly reduced on CU as compared to AN69. Onto CU, cells appeared to secrete a thick extracellular matrix and to be much more adherent than onto AN69. Moreover, on CU, cells were highly charged in melanin while they remained mostly uncolored onto AN69. These observations suggest cell differentiation on CU (Figure 3.9).

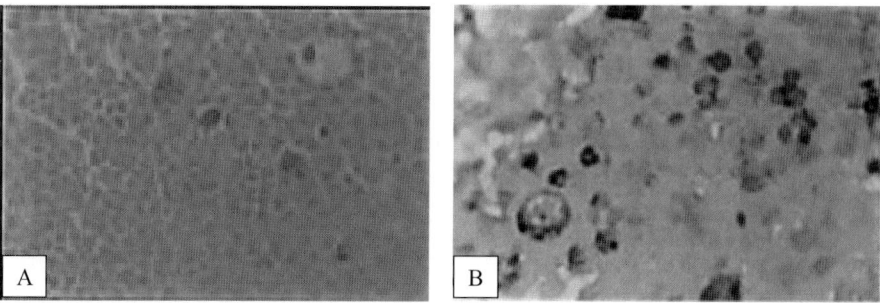

Figure 3.9. Melanin staining using Fontana-Masson technique of B16F10 explants cultivated for 2 weeks on AN69 (A) and CU (B). Explants cultivated on AN69 were melanin free, by contrast, on CU, explants were loaded in melanin (magnification × 500)

3.6 Conclusions

The whole of the results are summarized in Table 3.1. CU, which adsorbs poorly serum adhesive proteins, appears to reduce cell proliferation, increase total protein synthesis, and induce programmed cell death and/or differentiation.

CU and several other hemodialysis membranes are manufactured from cellulose. Modified cellulose membranes compared with synthetic membranes have been shown to induce higher C3a concentrations and superior leukopenia rate [32], to increase protein oxidation and systemic inflammation [33], and to enhance leukocyte activation [34], in hemodialysis patients. An increase in levels of adhesion molecules in the course of a single hemodialysis session has been demonstrated in CU patients [35]. By elsewhere, CU dialyzers could lead to a significantly greater secretion of TNF-alpha [36]. During hemodialysis, circulating mononuclear cells can be activated to different levels depending on the membrane used. This activation may generate an apoptotic process. Carracedo *et al.* [37] evidenced aggregation and apoptosis of circulating human mononuclear cells incubated in contact with CU membrane for 48 h. Our *in vitro* results gained from various cell types sustain these clinical observations.

The understanding of cell-surface interactions through the adsorbed serum adhesive proteins cannot be ignored for biomaterials improvement and new

Table 3.1. Comparative behavior of Swiss 3T3 fibroblasts (short-term cultures) and melanoma explants (long-term cultures) on CU and AN69 or PS used as controls

Cell strain behavior studied	Cuprophan low adsorption of serum adhesive proteins	AN69 or PS high adsorption of serum adhesive proteins
3T3 Cell morphology	Round and aggregated	Spread
3T3 cAMP production (45 min)	+ + +	+
3T3 RhoA activation (45 min)	+	+ + +
3T3 Cx 43 organization (45–180 min)	+ + +	+
3T3 Gap junction communication (90 min)	+ + +	
Proliferation 3T3: 24 and 48 h, B16F10 explant: 2 weeks	+	+ + +
Synthesis 3T3: protein synthesis (48 h), B16F10: explant melanogenesis (2 weeks)	+ + +	+
3T3 Apoptosis (24 and 48 h)	+ + +	+
B16F10 explant cell migration (2 weeks)	+	+ + +
B16F10 explant cell adhesion (2 weeks)	+ + +	+

developments in tissue engineering. CU may be considered as a precious tool to study *in vitro* biochemical mechanisms regulating cell functions. Its surface properties may open fields of research aiming to evaluate normal and cancer cell capability to differentiate and to die.

References

1. Chen, C. S.; Mrksich, M.; Huang, S.; Whitesides, G. M.; Ingber, D. E. Geometric control of cell life and death. *Science* **1997**, *276*, 1425–1428.
2. Dike, L. E.; Chen, C. S.; Mrksich, M.; Tien, J.; Whitesides, G. M.; Ingber, D. E. Geometric control of switching between growth, apoptosis, and differentiation during angiogenesis using micropatterned substrates. *In Vitro Cell Dev. Biol. Anim.* **1999**, *35*, 441–448.
3. Boudreau, N. J.; Jones, P. L. Extracellular matrix and integrin signaling: the shape of things to come. *Biochem. J.* **1999**, *339*, 481–488.
4. Faucheux, N.; Haye, B.; Nagel, M. D. Activation of the cyclic AMP pathway in cells adhering to biomaterials: Regulation by vitronectin- and fibronectin-integrin binding. *Biomaterials* **2000**, *21*(10), 1031–1038.
5. Hotchin, N. A.; Hall, A. Regulation of the actin cytoskeleton, integrins and cell growth by the Rho family of small GTPases. *Cancer Surv.* **1996**, *27*, 311–322.
6. Burridge, K.; Wennerberg, K. Rho and Rac take center stage. *Cell* **2004**, *116* (2), 167–179.
7. Ridley, A. J.; Hall, A. The small GTP-binding protein rho regulates the assembly of focal adhesions and actin stress fibers in response to growth factors. *Cell* **1992**, *70* (3), 389–399.
8. Barry, S. T.; Flinn, H. M.; Humphries, M. J.; Critchley, D. R.; Ridley, A. J. Requirement for Rho in integrin signalling. *Cell Adhes. Commun.* **1997**, *4*(6), 387–398.
9. van Kooten, T. G.; Spijker, H. T.; Busscher, H. J. Plasma-treated polystyrene surfaces: Model surfaces for studying cell–biomaterial interactions. *Biomaterials* **2004**, *25*(10), 1735–1747.
10. Faucheux, N.; Warocquier-Clerout, R.; Duval, J. L.; Haye, B.; Nagel, M. D. cAMP levels in cells attached to AN69 and Cuprophan: cAMP dependence of cell aggregation and the influence of serum. *Biomaterials* **1999**, *20*(2), 159–165.
11. Faucheux, N.; Nagel, M.-D. Cyclic AMP-dependent aggregation of Swiss 3T3 cells on a cellulose substratum (Cuprophan) and decreased cell membrane Rho A. *Biomaterials* **2002**, *23*(11), 2295–2301.
12. Lo, C. W. The role of gap junction membrane channels in development. *J. Bioenerg. Biomembr.* **1996**, *28*(4), 379–385.
13. Loewenstein, W. R. Junctional intercellular communication and the control of growth. *Biochem. Biophys. Acta* **1979**, *560*(1), 1–65.
14. Simon, A. M.; Goodenough, D. A. Diverse functions of vertebrate gap junctions. *Trends Cell Biol.* **1998**, *8*(12), 477–483.
15. Moorby, C. D.; Gherardi, E. Expression of a Cx43 deletion mutant in 3T3 A31 fibroblasts prevents PDGF-induced inhibition of cell communication and suppresses cell growth. *Exp. Cell Res.* **1999**, *249*(2), 367–376.
16. Evans, W. H.; Ahmad, S.; Diez, J.; George, C. H.; Kendall, J. M.; Martin, P. E. Trafficking pathways leading to the formation of gap junctions. *Novartis Found Symp.* **1999**, *219*, 44–54.
17. Yeager, M.; Unger, V. M.; Falk, M. M. Synthesis, assembly and structure of gap junction intercellular channels. *Curr. Opin. Struct. Biol.* **1998**, *8*(4), 517–524.

18. Faucheux, N.; Dufresne, M.; Nagel, M. D. Organisation of cyclic AMP-dependent connexin 43 in swiss 3T3 cells attached to a cellulose substratum. *Biomaterials* **2002**, *23*(2), 413–421.

19. Faucheux, N.; Zahm, J. M.; Bonnet, N.; Legeay, G.; Nagel, M. D. Gap junction communication between cells aggregated on a cellulose-coated polystyrene: Influence of connexin 43 phosphorylation. *Biomaterials* **2004**, *25*, 2501–2506.

20. Lampugnani, M. G.; Giorgi, M.; Gaboli, M.; Dejana, E.; Marchisio, P. C. Endothelial cell motility, integrin receptor clustering, and microfilament organization are inhibited by agents that increase intracellular cAMP. *Lab Invest.* **1990**, *63*(4), 521–531.

21. Schoenwaelder, S. M.; Burridge, K. Bidirectional signaling between the cytoskeleton and integrins. *Curr. Opin. Cell Biol.* **1999**, *11*(2), 274–286.

22. Han, J. D.; Rubin, C. S. Regulation of cytoskeleton organization and paxillin dephosphorylation by cAMP. Studies on murine Y1 adrenal cells. *Biol. Chem.* **1996**, *271*(46), 29211–29215.

23. Faucheux, N.; Correze, C.; Haye, B.; Nagel, M. D. Accumulation of cyclic AMP in swiss 3T3 cells adhering to a cellulose biomaterial substratum through interaction with adenylyl cyclase. *Biomaterials* **2001**, *22*, 2993–2998.

24. Homburg, C. H.; de Haas, M.; von dem Borne, A. E.; Verhoeven, A. J.; Reutelingsperger, C. P.; Roos, D. Human neutrophils lose their surface Fc gamma RIII and acquire Annexin V binding sites during apoptosis in vitro. *Blood* **1995**, *85*(2), 532–540.

25. Verhoven, B.; Schlegel, R. A.; Williamson, P. Mechanisms of phosphatidylserine exposure, a phagocyte recognition signal, on apoptotic T lymphocytes. *J. Exp. Med.* **1995**, *182*(5), 1597–1601.

26. Vermes, I.; Haanen, C.; Steffens-Nakken, H.; Reutelingsperger, C. A novel assay for apoptosis. Flow cytometric detection of phosphatidylserine expression on early apoptotic cells using fluorescein labelled Annexin V. *J. Immunol. Methods* **1995**, *184*(1), 39–51.

27. Earnshaw, W. C.; Martins, L. M.; Kaufmann, S. H. Mammalian caspases: Structure, activation, substrates, and functions during apoptosis. *Annu. Rev. Biochem.* **1999**, *68*, 383–424.

28. Desagher, S.; Martinou, J. C. Mitochondria as the central control point of apoptosis. *Trends Cell Biol.* **2000**, *10*(9), 369–377.

29. Hengartner, M. O. The biochemistry of apoptosis. *Nature* **2000**, *407*(6805), 770–776.

30. Gekas, J.; Hindie, M.; Faucheux, N.; David, B.; Lanvin, O.; Maziere, C.; Fuentes, V.; Gouilleux-Gruart, V.; Maziere, J. C.; Lassoued, K.; Nagel, M. D. The inhibition of cell spreading on a cellulose substrate (Cuprophan) induces an apoptotic process via a mitochondria-dependent pathway. *FEBS Lett.* **2004**, *563*, 103–107.

31. Duval, J. L.; Faucheux, N.; Warocquier-Clerout, R.; Nagel, M. D. Behavior of melanoma cells in cell and organ cultures: Use of biomaterials to activate cells. *Cells Mater.* **1999**, *9*(1), 31–42.

32. Germin, P. D. Comparison of biocompatibility of hemophane, cellulose diacetate and acrylonitrile membranes in hemodialysis. *Acta Med. Croat.* **2004**, *58*(1), 31–36.

33. Walker, R. J.; Sutherland, W. H.; De Jong, S. A. Effect of changing from a cellulose acetate to a polysulphone dialysis membrane on protein oxidation and inflammation markers. *Clin. Nephrol.* **2004**, *61*(3), 198–206.

34. Hernandez, M. R.; Galan, A. M.; Cases, A.; Lopez-Pedret, J.; Pereira, A.; Tonda, R.; Bozzo, J.; Escolar, G.; Ordinas, A. Biocompatibility of cellulosic and synthetic membranes assessed by leukocyte activation. *Am. J. Nephrol.* **2004**, *24*(2), 235–241.

35. Musial, K.; Zwolinska, D.; Polak-Jonkisz, D.; Berny, U.; Szprynger, K.; Szczepanska, M. Soluble adhesion molecules in children and young adults on chronic hemodialysis. *Pediatr. Nephrol.* **2004**, *19*(3), 332–336.

36. Hoffmann, U.; Fischereder, M.; Marx, M.; Schweda, F.; Lang, B.; Straub, R. H.; Kramer, B. K. Induction of cytokines and adhesion molecules in stable hemodialysis patients: Is there an effect of membrane material? *Am. J. Nephrol.* **2003**, *23*(6), 442–447.

37. Carracedo, J.; Ramirez, R.; Pintado, O.; Gomez-Villamandos, J. C.; Martin-Malo, A.; Rodriguez, M.; Aljama, P. Cell aggregation and apoptosis induced by hemodialysis membranes. *J. Am. Soc. Nephrol.* **1995**, *6*(6), 1586–1591.

Chapter 4

COTTON AND PROTEIN INTERACTIONS

Steven C. Goheen[1], J. Vincent Edwards (USDA)[2], Alfred Rayburn[1],
Kari Gaither[1], and Nathan Castro[1]

[1]*Battelle Northwest, Richland, Washington 99352, U.S.A.*
[2]*USDA/ARS, SRRC, New Orleans, LA 70124, U.S.A.*

Abstract

The adsorbent properties of important wound fluid proteins and cotton cellulose are
reviewed. This review focuses on the adsorption of albumin to cotton-based wound dressings and some chemically modified derivatives targeted for chronic wounds. Adsorption of
elastase in the presence of albumin was examined as a model to understand the interactive
properties of these wound fluid components with cotton fibers. In the chronic non-healing
wound, elastase appears to be over-expressed, and it digests tissue and growth factors, interfering with the normal healing process. Albumin is the most prevalent protein in wound
fluid, and in highly to moderately exudative wounds, it may bind significantly to the fibers
of wound dressings. Thus, the relative binding properties of both elastase and albumin to
wound dressing fibers are of interest in the design of more effective wound dressings. The
present work examines the binding of albumin to two different derivatives of cotton, and
quantifies the elastase binding to the same derivatives following exposure of albumin to the
fiber surface. An HPLC adsorption technique was employed coupled with a colorimetric
enzyme assay to quantify the relative binding properties of albumin and elastase to cotton.
The results of wound protein binding are discussed in relation to the porosity and surface
chemistry interactions of cotton and wound proteins. Studies are directed to understanding
the implications of protein adsorption phenomena in terms of fiber-protein models that have
implications for rationally designing dressings for chronic wounds.

4.1 Background

4.1.1 Understanding cotton fiber-protein interactions through chromatography

The general phenomenon associated with protein adsorption to solids has
been previously investigated [1]. One of the most commonly studied substrates

49

for protein adsorption in aqueous systems is cellulose and its derivatives [2, 3]. Cellulose is an excellent substrate for protein purification processes because, like the surface of soluble proteins, it is hydrophilic and typically non-denaturing. Raw cotton is mostly cellulose, but there are impurities, comprised primarily of lignin, waxes and fats, and proteins. Processed cotton has lower impurity concentrations. Liquid chromatography is a useful method of studying protein adsorption due to the process of differential adsorptive behavior of the sample components between the mobile and stationary phases [4–11]. We consider here the use of cotton as a liquid chromatography stationary phase. Considerable work has been done on the binding of individual amino acids to cellulose [12] and the binding of peptides and polyamino acids to cotton have been examined [13] as well. The unique primary, secondary, and tertiary structure of proteins allows them to bind a substrate, incorporate into a membrane, or transport a molecule among many other functions in nature [14]. Likewise, protein-fiber surface models may be created based on the unique binding properties of a protein and the properties of the fiber surface it binds to.

Protein adsorption has been extensively investigated with other protein-surface systems using adsorption isotherms [4, 15, 16]. The molecular interactions attributed to protein layering are hydrophobic, van der Waals and hydrogen bonding. These forces sometimes result in conformational changes thought to be responsible for the stronger binding of proteins that occurs over time [4–11, 15, 17].

We have applied the use of an inverse liquid chromatography technique [4, 9–11] to examine the protein sorption properties of cellulose [4]. It is termed 'inverse chromatography' because we are often more interested in the binding properties of the substrate than the solvated protein. In this method, a sorbent is loaded into columns and standard proteins are eluted over the substrate. Elution time and recoveries are both measured parameters that reveal important surface characteristics of the cotton-protein interactions [4]. The use of inverse chromatography is similar to the approach of studying protein binding to stationary substrates with high performance liquid chromatography (HPLC) [6–8]. HPLC has been used to quantify protein adsorption by careful calibration and peak integration. As a chromatographic method it can provide relative protein-substrate sorption strength by adjusting the eluting solvent or mobile phase. Protein losses from affinity or unfolding processes can also be determined using HPLC by a similar technique [4, 9–11].

When soluble proteins elute through any stationary phase media there are two notable sorption characteristics. These are the recovery and retention time of the protein. Recovery is a property that frequently depends on the number of collisions or interactions of the protein across the surface and the chemical affinity between the sorbent and the protein. Those proteins that are not recovered are bound strongly to the surface. The second characteristic is the retention time of the eluted proteins. Retention time is related to weak interfacial binding. In gradient elution retained sorbents bind one or more layers of proteins at the front end of the gradient, forming a partial layer, and then elute later in the gradient presumably with no overall loss of protein, or a slight loss with subsequent injections. However, we have shown that losses are common and appear to be correlated with residence time, sorbent chemistry, and protein molecular weight [9].

4.1.2 Albumin and elastase interaction with cotton fibers for design of improved wound dressings

Albumins are water soluble, acidic, globular proteins that are easily crystallized. Serum albumin (MW 67,500, and isoelectric point of 4.9) by virtue of being 55–60% of the protein material in blood serum is a major transport and compositional protein of early acute wounds [18]. Human neutrophil elastase (HNE; MW: 27,500; isoelectric point: 9.5) is an enzyme that catalyzes the cleavage of elastin at the peptide bond adjacent to a non-aromatic, hydrophobic amino acid. HNE has a broad substrate range which accounts for the rapid degradation of growth factors and extracellular matrix proteins under chronic wound conditions. Approximately a picogram of elastase resides in each polymorphonuclear leukocyte or human neutrophil cell, and there are, approximately, an average of one million neutrophils circulating in the body. Human neutrophils provide a line of inflammatory defense against microbes and the clearing of cellular debris when a wound first occurs. However, when the chronic wound becomes arrested in the inflammatory stage, an overabundant supply of neutrophils results in the release of dangerous levels of HNE that account for deleterious protein breakdown. Our interest in serum albumin and elastase interactions with cotton is related to wound healing. Serum albumin is the most concentrated protein found in wound fluid. While there are many other proteins in wound fluid besides albumin, elastase control in the chronic wound has been a target of cotton wound dressing design. The question of whether albumin and elastase compete for binding sites on cotton gauze becomes relevant for dressing design due to capacity issues that require a robust fiber that will continue to remove excess elastase from the chronic wound for a sustained period between dressing changes. This is an important concern because it is anticipated that a wound dressing can play an important role in the healing process with regards to elastase and the role of elastase in the pathophysiology of the chronic wound [19]. Previous studies to sequester and inhibit high levels of unwanted elastase activity in the chronic wound have involved the application of elastase peptide recognition sequences to fibers of wound dressings [20]. These elastase substrate peptides included the covalent linking of the Val-Pro-Val recognition sequence of elastase to cotton cellulose [20, 21]. Cotton is an ideal material for wound dressings because it is highly hydrophilic, absorbent, and inexpensive. In this study we examine the effect of employing modified cellulose that can be prepared more economically for regulation of elastase activity in the chronic wound [22].

An understanding of the surface binding characteristics of both albumin and HNE is vital to the design of wound dressing fibers. We have previously observed that serum albumin, with a net charge (outside its isolectric point) will bind either positively, or negatively charged surfaces [11]. This surface binding phenomenon may be due to a cluster of exposed negatively charged amino acids on the protein's surface. However, under extreme conditions, if the pH was low enough to protonate all exposed amino acid side chains, electrostatic adsorption to a positively charged surface would probably not occur. We more closely examine this binding phenomenon here with cotton fibers.

Solid materials are prone to deforming large proteins or other macro-molecules when they adsorb them [9]. The deformation of a protein upon contact with a solid surface can be so severe as to cause the protein to unfold. Upon unfolding a protein exposes additional sorption sites that can become involved in strong irreversible binding to surfaces [4]. This ability to induce deformation will be discussed here in relation to wound dressing design. We have previously shown that surfaces act like catalysts for protein deforma-tion and unfolding [4]. Inverse liquid chromatography is used in this study to examine cotton for its relative binding properties and potential unfolding of albumin.

This chapter focuses on protein adsorption to cotton in relation to wound healing and other biomedical applications. These are mostly examples designed to stimulate new ideas for future use. The reader is encouraged to apply the basic concepts of this chapter toward additional work to help solve the numerous challenges related to protein adsorption to fibers.

4.2 Materials and methods

4.2.1 Materials

4.2.1.1 Chemicals

Water was deionized and processed through a Millipore MilliQ Plus system such that the resistance of the water was greater than 18.2 ohms. Trizma (Tris[hydroxymethyl]aminomethane), bovine serum albumin (BSA), porcine pancreatic elastase, and substrate (N-(methoxysuccinyl)-ala-ala-pro-val 4-nitroanilide) were obtained from Sigma (St. Louis, MO). The Bradford dye reagent was obtained from Bio-Rad Laboratories (Hercules, CA). All reagents were of the highest purity. The cotton used was pre-ground to 80-mesh size using a Wiley Mill. Fiber fragments from both samples were isolated using an 80-mesh screen as described [21]. Cotton derivatives consisted of carboxymethylated cellulose (CMC) and an underivatized sample. The cotton samples were prepared as described previously [5, 22]. The CMC derivative was prepared as described [20].

4.2.1.2 Equipment

HPLC: Quantitation was performed using high performance liquid chromatography (HPLC) with a BioRad Series 500 HRLC and a Bio-Rad column heater (Bio-Rad Laboratories, Hercules, CA). The detector was a Gilson 118 UV/Vis unit that was set to 280 nm for the detection of albumin (or elastase). The flow rate was 1.0 mL/min. The eluents used are shown in Table 4.1. Calibrations were carried out using a low-dead volume (LDV) connector in place of a column. An LDV connector is a fitting with a small flow volume that can be connected in place of any HPLC column. The only proteins studied were BSA and elastase. Calibration curves were generated for each protein.

Spectrophotometer: A Cole Parmer 1200 spectrophotometer was used for enzyme activity measurements. The wavelength was kept constant at 410 nm for enzyme kinetic measurements. Proteins were assayed using HPLC as described.

Optical Micrograph: Samples were placed on a glass slide then viewed using reflected light followed with the placement of a glass cover slip over the fibers to attempt to bring them into the same focal plane. The samples

Table 4.1. Peak areas of BSA in the HPLC system for each individual buffer system

Buffer conditions effect on BSA				
Buffer	Tris	NaCl	pH	Peak area*
1	5 mM	0	9.0	1.66 ± 0.039
2	5 mM	0	7.4	1.66 ± 0.035
3	5 mM	0.15 M	9.0	1.62 ± 0.079
4	5 mM	0.15 M	7.4	1.59 ± 0.077

*Areas given are $\times 10^6$ area counts/second. The difference in peak areas between buffer 1 and 4 is less than 5%.

were then examined on an Olympus BX 60 light microscope (Olympus model BX60) at different magnifications and digital light micrographs were taken of the sample using a Kodak DC290 digital camera.

Scanning Electron Microscopy (SEM):   Both cotton fiber samples were prepared for viewing in a field emission scanning electron microscope in the same manner. A 13 mm SEM stub was cleaned and a small piece of graphite tape was placed on the top of the stub. Then a small amount of each cotton fiber sample was adhered to the graphite tape. The material that had not adhered was removed with a small discharge of air from a filtered air source. The sample was then allowed to sit in a vacuum of the Edwards S150B Sputter Coater for 1 hour to allow for the small particles to settle on the graphite tape. After the particles had settled, the samples were sputter coated with carbon for 10 seconds. The samples were removed from the sputter coater and placed in a N_2 fed dry box for storage.

The samples were viewed on the LEO 982 FESEM (Carl Zeiss, Inc., Thornwood, NY 10594) with varying magnifications. The LEO 982 FESEM has resolution capabilities of 1.0 nm at 30 kv. High and low magnification images were taken of the particles with the majority of the images being taken of common structures found in the majority of the samples. The samples were then removed from the FESEM and placed back in the dry box for storage and further examination.

4.2.2 Methods

4.2.2.1 Protein Quantitation

Four buffers were created in the pH range (7–9) of Trizma® salt to determine variability in absorbance readings over the operating range of the experiment. The buffer conditions can be seen in Table 4.1. Twenty microliters of 40 μg/mL BSA in 5 mM Tris, 0.15 M NaCl pH 7.4 was injected with one of the four buffers in Table 4.1 eluting at 1.0 mL/min. The same method was used in all four buffer systems. Pumps were operated remotely through the ValueChrom® software and absorbance was measured at 280 nm as described above. Peak areas were calculated with the built-in integrator. Each buffer was eluted through two dual-piston pumps operating at 50% each.

The amount of protein adsorbed was determined by subtracting the peak area of the protein (BSA) or enzyme activity (elastase) that eluded through the cotton column from that eluted through the LDV connector. Elastase activities were determined by adding the activity of fractions collected downstream from the detector. Three 0.5 mL eluted fractions were collected, analyzed for enzyme activity, and totals were added to obtain the eluted activity. The dynamic range of this method was verified by determining the linearity of elastase activities of diluted enzyme.

Albumin was examined through subsequent injections of BSA in 0.15M NaCl, 5 mM Tris, pH 7.4 until the absorbance values were measurable and had stabilized to a constant value. Elastase was injected in a similar manner, but only after the cotton column had been pre-conditioned by injecting repeated injections of BSA as described above. Several injections of elastase were also put through the LDV connector. In all cases, the amount of elastase injected was nearly identical to the amount eluted.

4.2.2.2 *Elastase assay*

Elastase activity was measured using the substrate (N-(methoxysuccinyl)-ala-ala-pro-val 4-nitroanilide) and measuring the rate of change in absorbance at 410 nm. This method has been described previously [21]. A 600 micro-molar solution of substrate was solubilized in dimethylsulfoxide (DMSO) and 100 microliters of this substrate solution was added to initiate the enzyme re-action. Assays were conducted in pH 7.4 buffer containing 5 mM Tris, 0.15 M NaCl, and the release of p-nitroaniline was measured at 410 nm from the enzy-matic hydrolysis of the substrate. The initial slope of the increasing absorbance at 410 nm was used to determine K values.

The procedures used to determine the activity of elastase in the presence of albumin that had been bound to CMC and underivatized cotton samples were identical. A 570 μg/mL elastase in 0.2 M NaCl solution was used. Elastase samples were injected into HPLC columns that had been conditioned with BSA as described above. The activity of each eluted fraction resulting from a single injection was determined and if several adjacent samples were active, the totals were added. Usually, only three consecutive 0.5 mL fractions contained measurable enzyme activity.

4.3 Results and discussion

4.3.1 Albumin absorbance and quantitation

Albumin binding was examined on CMC and underivatized cotton columns. Buffer conditions had no effect on BSA absorbance. Table 4.1 displays the lack of effect of buffer conditions on BSA absorbance (and peak area) at 280 nm.

Table 4.1 Covers a wide pH range of the Tris buffer. It was not anticipated that pH or [NaCl] would influence the peak area measurements, but since CMC cotton can act as an ion exchanger, shifts in pH and salt concentration are possible. Documenting the stability of the peak area measurements adds confidence that any change in peak area values represents an absolute rather than perceived loss in protein.

4.3.2 Microscopy

Optical and scanning electron micrographs of the cotton samples are shown in Figures 4.1 and 4.2. The fibrous structure of the samples derived from −80 mesh pulverized cotton gauze is evident. The fibers of the samples pulverized for use as a chromatography stationary phase were estimated to average $320 \times 18 \times 6$ μm in size. CMC and underivatized cotton fibers appeared similar. Both the light and electron micrographs demonstrated a fiber morphology that was consistent with a slightly twisted structure with smooth surfaces. Determination of the cotton fiber dimensions was used to calculate surface area (see Table 4.2).

Calculations of the surface area of the fiber samples were conducted based on the dimensions of individual fibers and their density, 1.6 g/cm^3 [23, 24]. The average surface area and volume of the cotton fibers analyzed was estimated to be 15.6×10^3 μm^2 and 34.6×10^3 μm^3 respectively for an average fiber (Table 4.2). Therefore, the surface area of a gram of cotton is about 3×10^{11} μm^2 (Table 4.2). The dimensions of BSA ($60 \times 60 \times 100$ Å) used in estimating the surface area occupied by a single BSA molecule gave a value of either 36×10^{-6} or 6×10^{-5} μm^2 depending on the orientation of the sorbent molecule. When the

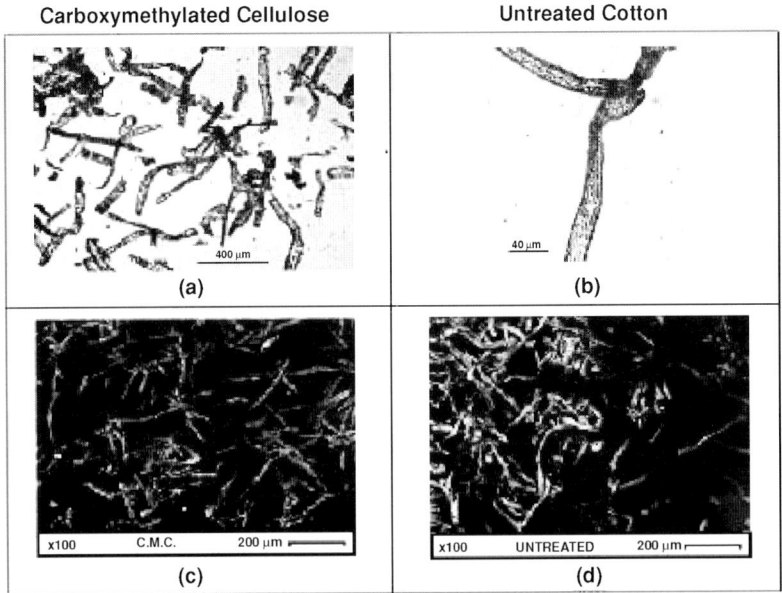

Figure 4.1. Light Micrographs [(a) and (b)] and Scanning Electron Micrographs (SEM) [(c) and (d)] of underivatized [(a) and (c)] and CMC derivatized [(b) and (d)] cotton samples

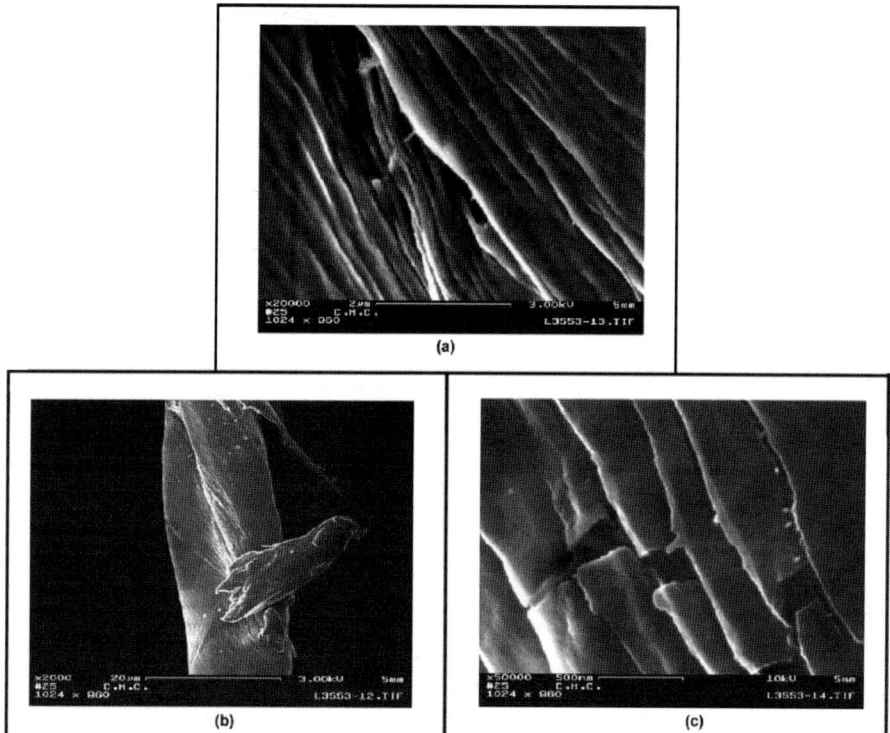

Figure 4.2. Scanning electron microscopy images of CMC cotton fibers. Images show cotton strand striation and fiber structure. Similar fields were magnified sequentially with the lowest magnification indicated in (a) and highest in (c). The horizontal breaks in (c) were from beam damage

surface area of the cotton fiber and albumin volume are factored together, the amount required to form a monolayer is approximately 0.5 mg BSA/g cotton.

Summarized in Figure 4.3 is the chromatographic profile of albumin quantitation following successive protein injections, elution and recovery of protein

Table 4.2. Fiber dimensions and cotton properties. The average fiber dimensions were estimated from micrographs like those shown in Figure 4.1. Cotton density is known [23, 24]. The surface area of cotton per gram was calculated from these values

Fiber dimensions and cotton properties	
Average fiber dimensions	$320 \times 18 \times 6$ μm
Density of cotton	1.6 g/mL
Surface Area of a fiber	15.6×10^3 μm^2
Volume per fiber	34.6×10^3 μm^3/fiber
Surface Area of cotton/g	2.82×10^{11} μm^3/g

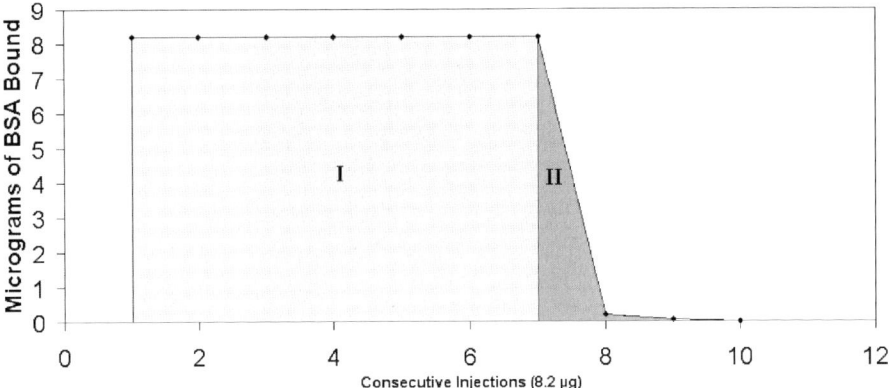

Figure 4.3. Adsorption of BSA on CMC treated cotton. Recoveries allowed us to calculate the amount bound with each injection. Based on this data, the saturation of CMC was attained at 65.6 µg BSA. All the injected BSA was recovered at injection 10. Two zones are shown. Zone I depicts the region where all the BSA was bound. In zone II, the CMC is becoming saturated with BSA

from the CMC column. Zone I in Figure 4.3 indicates all the injected BSA was being adsorbed to cotton. The amount of BSA bound (65.6 µg/g) to the CMC column was about 10% of that required to coat the surface of the cotton fibers with a monolayer of albumin assuming the shape of the albumin molecule does not deviate dramatically from its native conformation. The low binding of albumin may be due to adsorption to one or more impurities, or to the ends of the fibers. An alternative explanantion is binding to the rough surfaces (Figure 4.2) of the fibers.

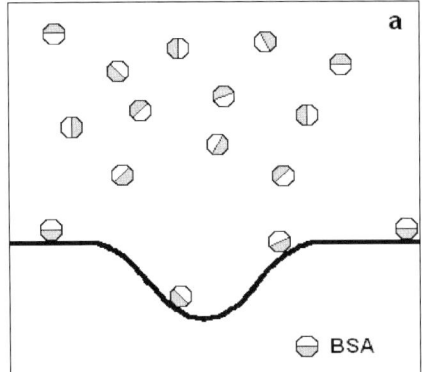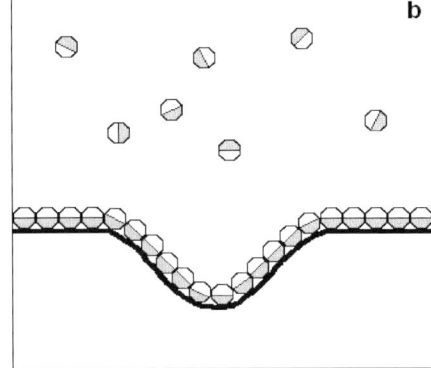

Figure 4.4. Graphical interpretation of BSA—cellulose fiber adsorption in CMC cotton. The adsorption mechanism remains unclear, but only a small percentage of the surface area sequestered BSA

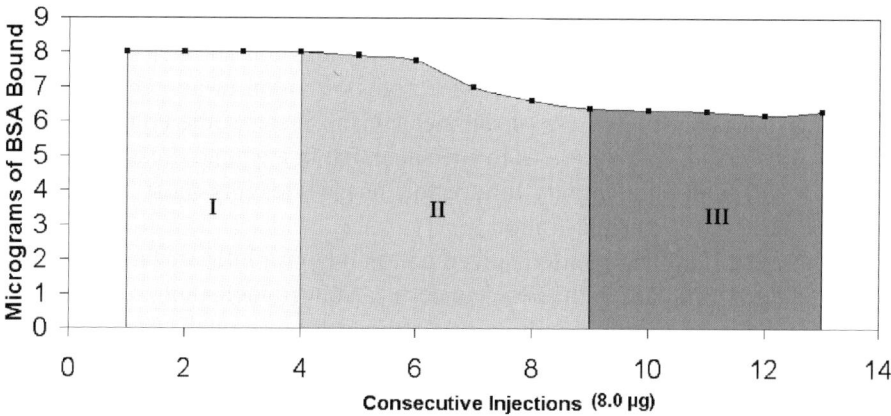

Figure 4.5. Adsorption of BSA to untreated (underivatized) cotton. Buffer containing 8 micrograms of BSA was injected 13 times through approximately 1.0 g underivatized cotton packed in an HPLC column. UV absorbance was monitored with each injection as described in Figure 4.3. Zones I and II correspond to similar zones in Figure 4.3. Zone III is a region in which BSA continues to bind with each subsequent injection

Albumin may deform when it adsorbs to cotton as we have shown for other proteins [2–4], but protein deformation induces a relatively small change in the fiber surface binding ratio. Zone II in Figure 4.3 can be viewed as the completion of the binding of BSA to the cotton surface.

The relative charge contribution to surface binding is an additional consideration in assessing the interaction of albumin to the CMC cotton. At pH 7.4, BSA has a net negative charge, as does the CMC surface. Both surfaces are hydrophilic which minimizes the possibility that BSA binds to the derivatized fiber surface. It is more likely that BSA adsorbed either to non-cellulose components of cotton, diffused into the ends of the fibers, adsorbed into pores, or became trapped between fibers in the packed column.

The chromatographic performance of underivatized (neutral, but hydrophilic) cotton was assessed to compare with albumin binding to CMC. In a similar manner to the analysis of albumin portrayed in Figure 4.3, albumin was bound to an underivatized cotton column by repetitive injections. Figure 4.5 depicts the binding profile of albumin to the column. As can be seen in Figure 4.5, zones II and I were similar to those in Figure 4.3, however in zone III sorption continues. This is apparent from the continued binding beyond that observed with the CMC cotton (Figure 4.3).

There are similarities and differences in the binding of albumin to the CMC and underivatized cotton. As can be seen in Figure 4.5, zones I and II were similar to those in Figure 4.3. However in Figure 4.5 the continued binding of albumin evident in zone III indicates sorption continues beyond that of CMC cotton.

An explanation for the observed behavior in Figure 4.5 is that CMC cotton repels anionic BSA whereas underivatized cotton does not. The lack of repulsion may allow more BSA to diffuse into the pores of cotton or adsorb weakly to the cotton surface. Regardless of the mechanism, most of the cotton surface did not bind BSA. This allows us to speculate that in a wound dressing, serum albumin will bind only slightly to a cotton surface, leaving most of the fiber available for protease sequestration.

In both the CMC and underivatized cotton a hydrophilic surface was able to bind a hydrophilic protein (BSA) under near-physiological conditions. The literature suggests that hydrophilic non-ionized (as in underivatized cotton) surfaces repel the adsorption of hydrophilic proteins (such as BSA) unless attracted by oppositely charged ionic regions of the surfaces [12]. And, it has been frequently postulated that proteins with a net charge will not adsorb to a surface of the same net charge [25]. Yet, a small amount of BSA clearly binds to CMC cotton.

We have also shown that albumin effectively binds to two different cotton derivatives (underivatized and CMC) in a slightly different manner. While further work is needed to elucidate the precise mechanism involved, it is clear that the CMC modification in cotton minimizes, but does not eliminate BSA binding. The mechanism of albumin binding to CMC may have important implications for the design of wound dressings. Since CMC is used as an occlusive wound dressing for heavily exudating wounds, understanding protein interaction on CMC is an important key in the future design of more effective dressings. However, all body fluids are far more complex that this simple system. Wound fluid also contains numerous other proteins, including elastase. Elastase is of particular interest in wound healing as discussed in the introduction. Thus, we examined the binding of elastase to modified fiber surfaces in the presence of albumin.

4.3.3 Elastase activity

One ultimate goal in conducting these experiments was to examine ways to inhibit elastase in environments similar to the chronic, non-healing wound using derivatized cotton gauze. Elastase can either be measured by weight or activity. And, it is possible that elastase activity is influenced by the environment. Thus, the concentration of elastase may not always be reflected by activity measurements alone. We used the cotton samples that had been saturated with BSA to bind additional elastase. For this experiment, we performed a simple comparison between the amount of elastase adsorption by weight (by optical absorbance) and enzymatic activity. While the CMC and underivatized cotton samples adsorbed different amounts of BSA, both had been exposed to excess BSA until the most aggressive binding region (in zone 1) was saturated. Since

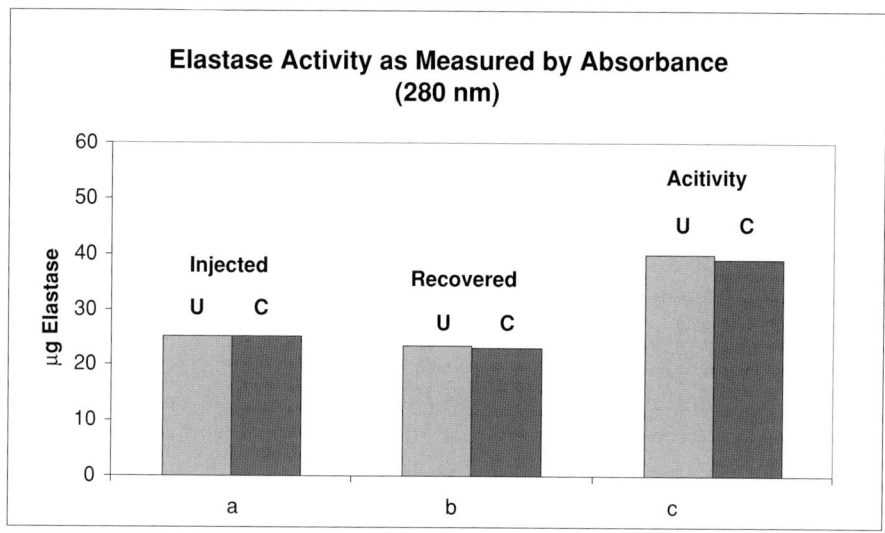

Figure 4.6. Elastase recoveries by mass (measured at 280 nm) and activity (measured by enzyme kinetics). Activity values were normalized to correspond to the amount of elastase injected. Underivatized cotton (U) and CMC cotton (C) are shown. (a) amount of injected Elastase, (b) amount of recovered elastase as measured by absorbance at 280 nm from the HPLC (cotton) column, (c) amount of elastase activity recovered from HPLC (cotton) column relative to the amount injected (see materials and methods)

the two cotton derivatives absorbed different amounts of BSA, they cannot be compared to one another. But, both forms can be examined for their ability to sequester elastase. Figure 4.6 shows these results.

When porcine pancreatic elastase was eluted through columns containing BSA bound to either the CMC or underivatized cotton, both the BSA-saturated cotton samples (underivatized and CMC cotton) adsorbed a small amount of elastase (Figures 4.6a and 4.6b). The CMC cotton bound slightly more presumably because elastase is cationic at pH 7.4 and CMC cotton has a net negative charge at that pH. The huge losses of protein as observed for BSA did not occur on either of the cotton materials, and even though less protein eluted in both cases, more elastase activity was recovered after exposure to either form of cotton when exposed to that representing the initial elastase injected into the column (Fig. 6c). This increased activity on both types of cotton may be due, in part, to an activation of elastase by BSA [5]. It may also be due to the removal by BSA or cotton of elastase inhibitors in the commercial enzyme preparation. However, we used HPLC and a reversed phase column to look for differences in the profiles of elastase exposed to cotton and unexposed elastase. We found no differences in these profiles (unpublished observation). It would be valuable to explore this apparent activation process further, especially since elastase activity is of concern in chronic non-healing wounds.

These results identify some new information regarding protein binding to cellulose. The two proteins that were selected were introduced to cotton in different ways. Serum albumin appears to bind sparingly to cotton whereas a smaller amount of elastase would bind to BSA-treated cotton. It made little difference whether elastase was introduced to BSA-treated cotton in the underivatized or CMC forms. This behavior suggests that the presence of albumin may minimize additional protein binding, however, each protein is likely to be unique in its binding properties. The displacement of BSA by elastase did not occur since the enzyme activity was preserved. However, displacement behavior of proteins on surfaces has been observed for other proteins [16].

We have only considered the cellulose molecule with and without protein coatings as sorption surfaces. Clearly, purified cotton (such as the cotton gauze used in this study) consists of very small concentrations of impurities. In particular, lignin, a significant component of raw cotton [26], has been shown to bind BSA above pH 4 [27]. We, therefore, suggest that BSA may have bound to lignin rather than cellulose in both forms of cotton, and that a separate mechanism was responsible for continued binding in underivatized cotton. And, since CMC and underivatized cotton were physically similar, but chemically different, we anticipate it was the chemical modifications of the cellulose or other impurities that caused these to behave differently.

Water plays a major role in protein adsorption [28], yet little has been mentioned in this chapter about its role in the binding of either elastase or BSA to cotton. The role of water can be imagined as protecting surfaces from protein binding. Water coats the surfaces of both the protein (if they are water-soluble) and sorbents (if they are hydrophilic) to insulate the two from a direct collision. Since the proteins that are soluble in wound fluid are hydrophilic and cotton is also hydrophilic, binding of these proteins to the gauze surface should be minimal. However, adsorption of BSA and to a lesser extent, elastase, clearly exists for both types of cotton studied here. Still the mechanism remains uncertain.

Natural biochemical processes can involve the adsorption and desorption properties of proteins as they perform their most important functions. For example, enzymes require close association with substrates to function as catalysts. Another example is membrane proteins that act as channels. They must have unique sorption characteristics to perform their natural function. It is possible that the adsorption behavior of both elastase and BSA to cellulose, provides key information to their *in vivo* function. For example, even though elastase is oppositely charged to that of BSA at pH 7.4, it appears to resist extensive adsorption. This is likely due to the hydration of both cotton and elastase. Albumin, on the other hand, when exposed to a like-charged adsorbed sparingly, but then repelled additional binding.

The results for this chapter were gathered using just a few simple examples. However, wound fluid is much more complex than two individual proteins in a simple buffer. It is not clear whether our results have physiological significance, or if our findings are just isolated phenomena. But, these results do reflect the complexity of seemingly simple sorption processes.

Other processes such as the sorption of dyes or stains, soil, etc. are likely to be just as complex as these simple examples. Dyes used to color fibers usually rely on covalent bonds to permanently attach them to the fiber [29–32]. However, unintentional soiling of fibers by macromolecules such as oils, proteins, and humic acids may involve processes similar to those identified here.

4.4 Conclusions

Protein adsorption to solid media is a widely studied field. Protein adsorption to cotton is not as actively investigated, but cellulose-protein interactions have been heavily studied especially in the chromatography literature. The present work was undertaken to examine the adsorption of two proteins, both of which are involved in wound healing. The adsorptive behavior of these two proteins, BSA and elastase, were investigated in different ways to model a wound situation. Albumin has adsorbed sparingly, and possibly to impurities such as lignin in the cotton matrix for CMC cotton, and continued to bind in the presence of underivatized cotton. Elastase did not bind as extensively to cotton that had been saturated with BSA, and elastase was activated after being exposed to the BSA-saturated cotton matrix. The activity of elastase was stimulated by both treated and underivatized cotton samples. Further studies elucidating the increase in elastase activity in the presence of albumin are ongoing. It is clear that more work is needed to fully understand not only these processes which relate to wound healing, but also to protein adsorption processes in general.

Acknowledgments

We thank the USDA for their financial support and for those in the Analytical Chemistry Group for assistance throughout this project. We thank Brian Pultz for supplying SEM and light micrograph images and preparation procedure for the analysis of the cotton samples studied. Andrew Mashchak and Esther Mintzer performed the HPLC analysis of elastase purity. We thank the Science and Engineering Education Department, the Office of Fellowship Programs, and Royace Aiken and Dale Johns of PNNL and the Department of Energy for additional financial support.

References

1. Missirlis, Y. F.; Lemm, W. (Eds). *Modern Aspects of Protein Adsorption on Biomaterials*. Kluwer Academic Publishers, Dordrecht, 1991, 288.

2. Tzanov, T.; Andreans, J.; Guebitz, G.; Cavaco-Paulo, A. Protein interactions in enzymatic processes in textiles. *Electronic J. Biotechnol.* 2003, 6(3), 146–154.

3. Ruckenstein, E.; Guo, W. Cellulose and glass fiber affinity membranes for the chromatographic separation of biomolecules. *Biotechnolo. Progress* 2004, 20, 13–25.

4. Goheen, S. C.; Gibbins, B. M.; Hilsenbeck, J. L.; Edwards, J. V. Retention, unfolding, and deformation of soluble proteins on solids. In: Edwards, V. E.; Vigo, T. L. (Eds.) *Bioactive Fibers and Polymers, Chap. 2*, American Chemical Society, Washington, DC, 2001, 20–34.

5. Edwards, J. V.; Yager, D.; Bopp, A.; Diegelmann, R. F.; Goheen, S.; Cohen, I. K. Design, preparaton and activity of cotton gauze for use in chronic wound research. In Edwards, J. V.; Vigo, T. L. (Eds.) *Bioactive Fibers and Polymers, Chap. 6*, American Chemical Society, Washington DC, 2001, 76–89.

6. Lu, X.-M.; Figueroa, A.; Karger, B. Intrinsic fluorescence and HPLC measurement of the surface dynamics of lysozome adsorbed on hydrophobic silica. *Journal of the American Chemical Society* 1988, 110, 1978–1979.

7. Kopaciewicz, W., Rounds, M.; Fausnaugh, J.; Regnier, F. Retention model for high performance ion exchange chromatography. *Journal of Chromatography* 1983, 226, 3–21.

8. Kopaciewicz, W.; Rounds, M. A.; Regnier, R. E. Stationary phase contributions to retention in high performance protein chromatography: Ligand density and mixed mode effects. *Journal of Chromatography* 1985, 318, 157–172.

9. Goheen, S. C.; Gibbins, B. M. Protein losses in ion exchange and hydrophobic interaction HPLC. *Journal of Chromatography A.* 2000, 890, 73–80.

10. Herbold, C. W.; Miller, J. H.; Goheen, S.C. Cytochrome c unfolding on an anionic surface. *Journal of Chromatography A.* 1999, 863, 137–146.

11. Goheen, S. C.; Hilsenbeck, J. L. High-performance ion-exchange chromatography and adsorption of plasma proteins. *Journal of Chromatography A.* 1988, 816, 89–96.

12. Kuhn, A. O.; Lederer, M. Adsorption chromatography on cellulose: II. Separations of aromatic amino acids, biogenic amines, alkaloids, dyes and phenols and determination of hydrophobic constants. *Journal of Chromatography A.* 1988, 440, 165–182.

13. Chirigos, M. A.; Papademetriou, V.; Bartocci, A.; Read, E.; Levy, H. B. Immune response modifying activity in mice of polyinosinic: Polycytidylic acid stabilized with poly-L-lysine, in carboxymethylcellulose [Poly-ICLC]. *International Journal of Immunopharmacology* 1981, 3(4), 329–337.

14. Goheen, S. C.; Gaither, K. G.; Rayburn, A. R. Nature and nanotechnology. In: Tsakalakos, T.; Ovid'ko, I. A.; Vasudevan, A. K. (Eds.) *Nanostructures: Synthesis, Functional Properties and Applications*, NATO Science Series II. Mathematics, Physics and Chemistry—Vol. 128, Kluwer Academic Press, Boston, MA, 2003, 117–137.

15. Chatelier, R. C.; Minton, A. P. 1996. Adsorption of globular proteins on locally planar surfaces: Models for the effect of excluded surface area and aggregation of adsorbed protein on adsorption equilibria. *Biophysical Journal* 1996, 71, 2367–2374.

16. Baszkin A.; Boissonnade, M. M. Competitive adsorption of albumin and fibrinogen at solution-air and solution-polyethylene interfaces. In situ measurements. In: Horbett, T. A.; Brash, J. L. (Eds.) *Proteins at Interface II, Chap. 15*, Amer. Chem. Soc., Washington, DC, 1995, 209–227.

17. Balasubramanian, V.; Grusin, N. K.; Bucher, R. W.; Turitto, V. T.; Slack, S. M. Residence-time dependent changes in fibrinogen adsorbed to polymeric biomaterials. *J. Biomedical Materials Research* 1999, 44(3), 253–260.

18. He, X. M.; Carter, D. C. Atomic structure and chemistry of human serum albumin. *Nature* 1992, 358, 209–215.

19. Enoch, S.; Price, P. Cellular, molecular and biochemical differences in the pathophysiology of healing between acute wounds, chronic wounds and wounds in the aged. *World Wide Wounds* 2004.

20. Edwards, J. V.; Batiste, S. L.; Gibbins, B. M.; Goheen, S. C. Synthesis and activity of NH2- and COOH-terminal elastase recognition sequences on cotton. *Journal of Peptide Research* 1999, 54, 536–543.

21. Edwards, J. V.; Eggleston, G.; Yager, D. R.; Cohen, I. K.; Diegelmann, R. F.; Bopp, A. F. Design, preparation and assessment of citrate-linked monosaccharide cellulose conjugates with elastase-lowering activity. *Carbohydrate Polymers* 2002, 50, 305–314.

22. Edwards, J. V.; Yager, D. R.; Cohen, I. K.; Diegelmann, R. F.; Montane, S.; Bertoniere, N.; Bopp, A. F. Modified cotton gauze dressings that selectively absorb neutrophil elastase activity in solution. *Wound Repair and Regeneration* 2001, 50–58.

23. Ward, K. *Chemistry and Chemical Technology of Cotton.* Interscience Publishers, Inc., New York, 1955.

24. Orr, R. S.; Weiss, L. C.; Moore, H. B.; Grant, J. N. Density of modified cottons determined with a gradient column. *Textile Research Journal* 1955, 25(7), 592–600.

25. Oscarrson, S. 1997. Factors affecting protein interaction at sorbent surfaces. *Journal of Chromatography B: Biomedical Science Applications* 1997, 699, 117–131.

26. Lynd, L. R.; Weimer, P. J.; van Zyl, W. H.; Pretorius, I. S. Microbial cellulose utilization: Fundamentals and biotechnology. *Microbial Mol. Biol. Rev.* 2002, 66(3), 506–577.

27. Zahedifar, M.; Castro, F. B.; Orskov, E. R. Effect of hydrolytic lignin on formation of protein-lignin complexes and protein degradation by rumen microbes. *Animal Feed Science and Tech.* 2002, 95, 83–92.

28. Bujnowski, A. M.; Pitt, W. G. Water structure around enkephalin near a PE surface: A molecular dynamics study. *Journal of Colloid and Interface Science* 1988, 203, 47–58.

29. George, B.; Govindaraj, M.; Ujiie, H.; Wood, D.; Freeman, H.; Hudson, S.; El-Shafei, A. Integratoin of fabric formation and coloration processes. National Textile Center Progress Report, 2003, 1–10.

30. United States Environmental Protection Agency. *Background Document for Identification and Listing of the Deferred Dye and Pigment Wastes*, 1999, 1–23.

31. Kim, T. K.; Yoon, S. H.; Son, Y. A. Effect of reactive anionic agent on dyeing of cellulosic fibers with a berberine colorant. *Dyes and Pigments* 2004, 60, 121–127.

32. Kim, T. K.; Son, Y. A. Effect of reactive anionic agent on dyeing of cellulosic fibers with a berberine colorant–Part 2: Anionic agent treatment and antimicrobial activity of a berberine dyeing. *Dyes and Pigments* 2005, 64, 85–89.

Chapter 5

ELECTROSPUN NANOFIBERS FROM BIOPOLYMERS AND THEIR BIOMEDICAL APPLICATIONS

Gisela Buschle-Diller, Andrew Hawkins, and Jared Cooper

Textile Engineering Department, Auburn University, AL 36849, U.S.A.

5.1 Introduction

Over the past decade electrospinning has been rediscovered as a process to generate ultrafine fibers, and a massive amount of research on the process and/or the product fibers has been published [1, 2, 3]. Electrospinning was originally pioneered by Formhals who filed for a patent on this process in 1934 [4]. Recent interest has been spurred by the fact that electrospun fibers have a high surface area-to-volume ratio. With such small diameter electrospun fibers enter the realm of nanomaterials. The medical relevance of nanomaterials has recently been reviewed [5]. Electrospun fibers are more exciting than most other nanofibers because their composition is highly diverse. Until recently, nanofibers have consisted largely of carbon nanotubes and other inorganic fibers. However, electrospun fibers can be fabricated of almost limitless materials from synthetic to natural polymers. Thus, the surface chemistry can be tailored to meet a large number of applications. The focus for electrospun products has so far included nonwovens for filtration [6], membranes for aerosol purification [7], thin coatings for defense and protection [8], and structures incorporated in composites [2]. Biomedical applications include more efficient wound healing and drug delivery devices, biocompatible scaffolding for tissue regeneration, bioerodable implant structures, and others [9, 10]. The purpose of this chapter is to briefly review some of the natural and synthetic biopolymers that can be electrospun into micro- and nanofibers and to show their value for future medical applications. Due to the enormous amount of research published

67

J. V. Edwards et al. (eds.), Modified Fibers with Medical and Specialty Applications, 67–80.

in recent years only a brief review will be presented here, with the scope of this article being to introduce the reader to this rapidly emerging field.

5.2 Principle of electrospinning

Conventional fiber formation techniques are not capable of producing fibers smaller in diameter than 2 μm. For example, a commercial polyester fiber typically has a diameter of approximately 10 μm. For most technical and textile applications larger fiber diameters in the range of 10–500 μm are sufficient and appropriate for their intended usage. However, in cases where interfacial characteristics and porosity are of decisive importance, electrospinning enables the production of fibers with diameters of less than 500 nm and thus with a high surface area-to-volume ratio. These fibers can easily be formed into nonwoven mats offering advantages over inorganic nanotubes and nanowires as well as other forms of nanostructured materials such as beads, films, or foams.

Electrospinning is similar to the electrospraying technique [6]. Basically any polymer that can be dissolved or melted at moderately high temperatures and without decomposition can be used for electrospinning. Fibers electrospun from solution are generally finer than fibers spun from melt due to the evaporation of the solvent in the former case.

The basic experimental set-up is presented in Figure 5.1. Electrospinning can be performed using either a horizontal or vertical geometry. For less viscous polymer spinning solutions the horizontal geometry is more advantageous. An

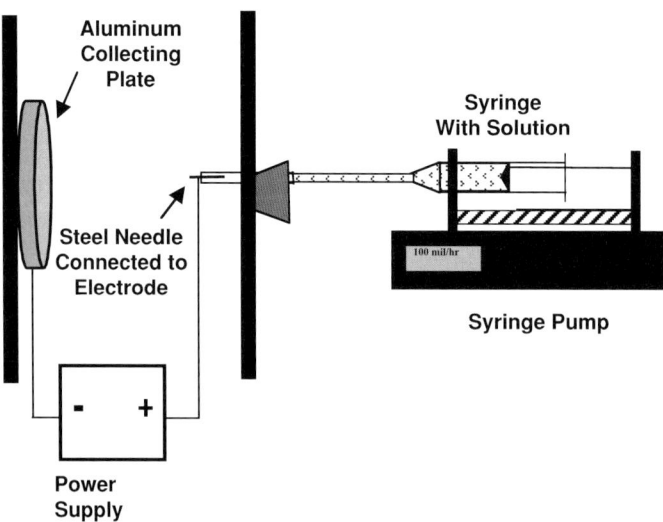

Figure 5.1. Schematic of the electrospinning apparatus

electric force, applied between the spinneret and a collector, draws the polymer jet toward the collection plate. The formation of nanoscale fibers is due to instabilities during spinning [6, 3], which cause the polymer jet to be elongated and possibly separated into smaller diameter fibers while whipping around or simply be stretched into finer fibers [3] before depositing on the target. Generally, the properties of the formed nanofiber web are determined by the polymer type and composition, the surface tension, viscosity, and conductivity of the spinning solution or melt, and the physical and geometrical parameters of the electrospinning, such as voltage and electrode distance. According to the model developed by the research group of Rutledge [11] the polymer jet reaches a terminal diameter which is determined by the characteristics of the polymer and its solvent, with the fluid's elastic properties being most vital for fiber formation.

The device used for electrospun biopolymers reported in this chapter involved a HV power supply from Gamma High Voltage Research with a range of 0–30 kV. A glass or plastic syringe with variable tip diameter was used as the spinneret. The product fibers were collected on different types of targets: scanning electron microscopic (SEM) stubs, rotating or stationary metal screens, knitted polyester tubing with a metal bar inside, paper, and glass, depending on the electrospun polymer. Details on polymer solution, specific solvent, voltage, electrode distance, etc. are given in ensuing sections.

5.3 Challenges of the process

Although conceptually a simple process, electrospinning has significant challenges. The major criticism of the electrospinning technique has been the comparatively slow, batch-wise production of the nanofibers. However, increasing commercialization of the process has led to technical advances; therefore the production rate should continually improve until an industrial scale process is more easily attained. For example, Donaldson Comp. [12] recently filed for a patent on electrospinning of nylon 6, 6,6, and 6,10 and blends for filter materials at a higher speed.

Electrospinning forms nanofibers as a product of instabilities of the polymer jet during spinning. This process causes the filament to stretch or possibly splice into even finer fibers. However, the whipping motion of the polymer jet randomizes the orientation of the nanofibers such that nonwoven mats are the simplest secondary structure to generate. Various different experimental designs have been constructed to improve the orientation and collection methods, as well as increase crystallinity and ultimately tensile strength by using a rotating drum [6, 1] or frame [2] as the collector. Charged rings have been proposed to help guide the jet stream. Electrospinning in a vacuum has also been explored [13].

Lee et al. [14] has identified numerous modifications regarding continuous as opposed to batch-wise fiber collection.

As a further challenge the web frequently contains fibers of varying diameters as well as individual fibers with varying thickness. In addition, small beads can form with the fibers. The formation of beads or fused crossover points is often due to delayed coagulation. This crisscross structure is undesired when fiber strand production is the goal of the electrospinning process. However, it can be useful as reinforcement of the nonwoven mat ensuring some stability to the fibrous web.

It is anticipated that electrospinning will continue to advance until the technical challenges will be overcome. In the meantime, it is likely that nanofibrous matted structures will be more fully explored.

5.4 Biopolymers for medical applications

For many biomedical applications, the slow production rate and web structure are not major disadvantages. This is especially true when the porosity and surface area properties are being exploited. Improvements in biomedical materials (implants, prosthetics, tissue replacement, and drug delivery devices) are continually in demand. Much of the current research is focused on natural materials for improved biocompatibility. Various physical and chemical properties define the potential biomedical use of a material. First, chemical reactions such as adsorption/recognition processes at the interface of the biomaterial influence biocompatibility. Also, physical properties, such as modulus of elasticity, need to match that of the neighboring tissue. Ideally, the device is truly integrated into the body's natural environment at a rate exactly equal to its being replaced by healthy tissue without causing an adverse reaction. In reality, distinction has to be made between those materials that are bioinert or cause no host response and no property changes over time, and those that biodegrade or bioresorb during which the implant decomposes in a natural, controlled manner with the degradation products being removed and replaced by the body in normal metabolic processes.

The physical compatibility of biomaterials includes variables such as structural integrity, strength, deformation resistance, fatigue properties, and modulus of elasticity. For prosthetic devices, carefully engineered metallic or ceramic biomaterials have adequate mechanical properties, wear, and corrosion resistance. However, metals and ceramics do not match both modulus and resiliency of living bone. Bone is continually undergoing fracture and repair processes whereas current synthetic materials do not have this property. Attempts to overcome this challenge have included making the surface or the entire material porous or biodegradable. Porous materials are used to encourage tissue growth into the prosthetic. Biodegradable materials are chosen so that the prosthetic will gradually disappear and be replaced by living tissue. To date, neither approach has been highly successful.

Biopolymers are better suited for applications that require flexibility, elasticity, and shapeability. Examples for biopolymer stets include wound dressings, drug release devices, soft tissue replacements, cardiovascular grafts, and sutures. Research in biodegradable polymers has increased dramatically over the past decade and good reviews are available in the literature (see e.g., Refs. [15, 10]).

Materials with high surface area and extended pores have a built-in scaffold for cell adhesion and cell in-growth. Porous coatings and modifications that render the implant surface rough and irregular encourage cell adhesion and favorable interactions with biological tissue. Examples for surface modifications include plasma treatments and grafting of either charged molecules, hydrophobic side-chains, or peptides to enhance cell attachment [16]. Electrospinning offers a method to produce high surface-to-volume ratio materials with extended porous systems easily. Besides the scaffolding itself being made from electrospun fibers, an interphase in the form of electrospun coatings that could adhere to a prosthetic device could be made as a transition region to the host tissue [2]. In the following sections a few more specific examples of potential biomedical applications of the electrospinning technology will be introduced.

5.4.1 Collagen

Collagen is the most abundant protein in the body of invertebrates and forms the principal structural framework of connective tissue such as ligaments, skin, tendons, and other organs. It is composed of a triple helical arrangement of amino acids with glycine at about every third position and stabilized by hydrogen bonding [17, 16]. A total of 19 collagen types have been established which differ in the composition of their individual chains within the triple helix.

Electrospinning of collagen for scaffolding and tissue engineering has been intensively studied over the past few years. Collagen can be isolated from various sources, however reprocessing of the protein into structures mimicking natural arrangements and suitable for scaffolding has been a major challenge. The collagen source largely determines the properties of the reprocessed product fibers. Extensive work has been performed by Bowlin, Simpson, and coworkers [18] primarily on electrospinning of collagen type I from calf skin. Comparisons were drawn to electrospinning of collagen types I and III from human placenta. A patent covering the technology has been filed recently by Simpson et al. [19].

Figure 5.2 shows fibrous webs of acid-soluble collagen type III formed by electrospinning of a suspension of 0.04 g/mL collagen in 1,1,1,3,3,3

(a)

(b)

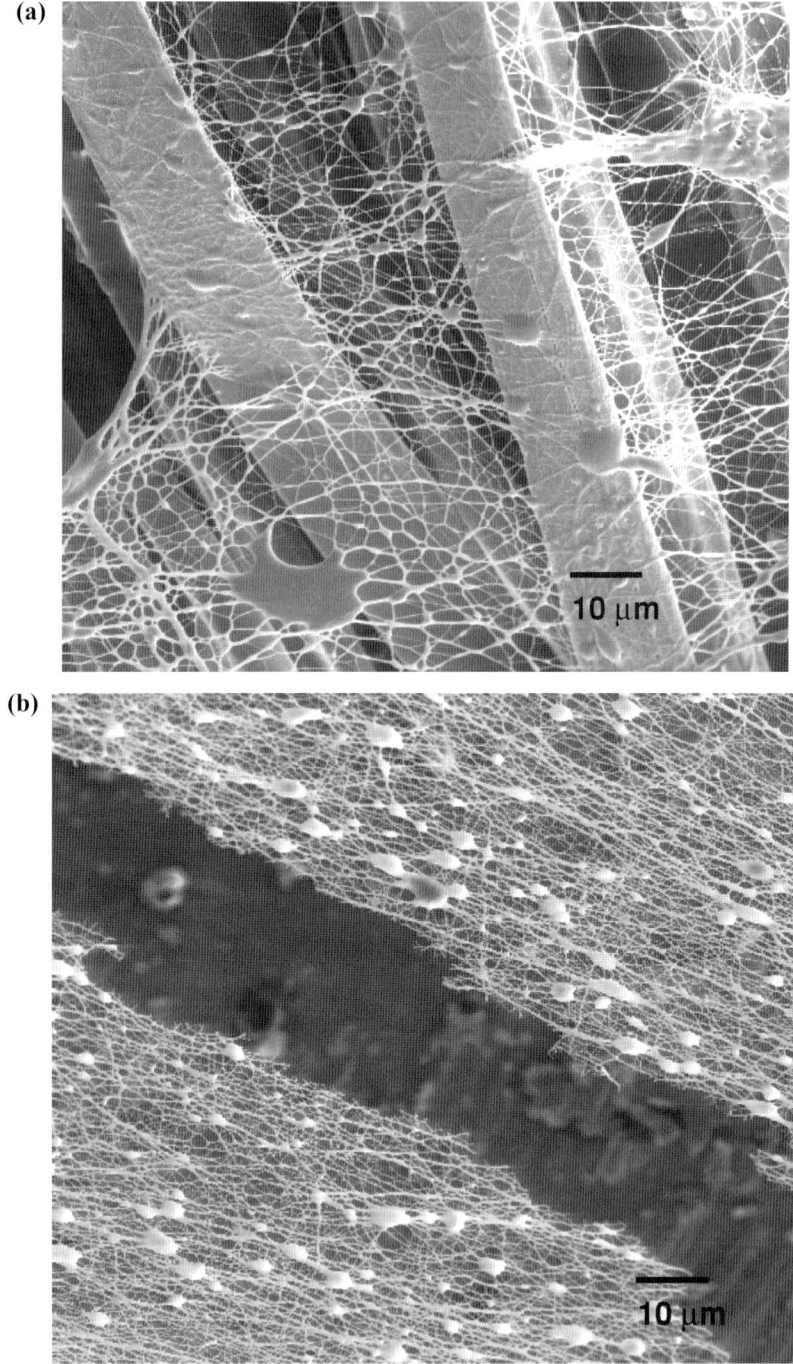

Figure 5.2. (a) Collagen spun on polyester scaffold (×500) and (b) a collagen web spun on a metal grid (×500)

hexafluoro-2-propanol (HFP) [18] in our lab. The voltage was 25 kV and the distance between electrodes 25 cm. The collagen web was deposited on different carriers such as a knitted polyester tubing mounted on a metal or glass core [20]. Fused fibers and small droplets within the web might be able to provide some mechanical stability to the sample. Such structures are advantageous for tissue engineering but might also affect the elasticity of the nonwoven structure. Thus, we have attempted to control the coagulation behavior of the fibers and the formation of beads by adjusting the temperature in the electrospinning region during the process.

Multilayered structures of electrospun collagen type I from bovine skin, styrenated gelatin, and segmented polyurethane as well as mixed component systems electrospun from polyurethane and poly(ethylene oxide) by simultaneous electrospinning have been reported by Kidoaki et al. [21]. Morphology and tensile properties of each component were investigated and evaluated in regard to improved functionality of scaffolding materials. The function of polyurethane was to provide elastomeric characteristics, while the gelatin fibers were tested for drug releasing and collagen for cell adhesion capabilities.

5.4.2 Poly(lactic acid), poly(glycolic acid), and poly(lactide-*co*-glycolide)

Poly(glycolic acid)   Poly(lactic acid)

Biocompatible and biodegradable, poly(lactic acid) (PLA), poly(glycolic acid) (PGA), and their copolymers are interesting materials for implants, sutures, and especially controlled drug release at high loading concentrations. The rate at which the drug is discharged into the biological system is determined by the degradation rate of the polymeric carrier. The degradation products, for instance lactic acid in the case of PLA decomposition, can easily be metabolized by the body.

Due to a fairly high melting temperature, PGA fiber formation has so far been limited to melt extrusion and drawing, producing fibers in the micrometer range. Boland et al. [22] could demonstrate that PGA can be electrospun from solution to produce nanofibers for tissue engineering.

Bognitzki et al. [23] used volatile solvents to create highly porous PLA fibers. While the fibers had a fairly large diameter in the micrometer range, regular pores developed of approximately 100 nm width and 250 nm length with orientation along the fiber axis. The authors noticed a rapid phase separation due to the evaporation of the solvent, followed by rapid solidification. It is believed that these processes combined to produce the observed phenomenon of fibers with large pores.

Zong et al. [24] investigated the effect of ions (salts) on the morphology of electrospun amorphous poly(D,L-lactic acid) and semicrystalline poly(L-lactic acid) fibers. The salts served as mimics for ionic drugs (e.g., Mefoxin®, cefoxitin sodium) to be included at a later stage.

Copolymerized products of PLA and PGA are available at various ratios of lactide (LA) and glycolide (GA) and exhibit different levels of crystallinity and mechanical properties [15]. Commercially available copolymers from 90:10 GA:LA have been used as sutures (Vicryl®). Of course, these threads are not in the nanofiber range.

Depending on the composition of PLGA, micro- or nanofibers can be produced by electrospinning. Figure 5.3 shows a scanning electron micrograph of PLGA nanofibers electrospun from melt at 220 °C in our lab.

Electrospun PLGA fibers from melt were straight and fairly large in diameter with an overall average of more than 1.5 μm. The smallest fibers (diameter 1.28 μm ($\pm$0.4)) could be obtained at a voltage of 20 kV and an electrode distance of 12–15 cm. The advantage of melt-electrospinning in this case is the option to fairly easily co-extrude PLGA together with a second melt-spinnable polymer to form bicomponent fibers with diameters in the low micrometer range (the results of these experiments will be presented elsewhere).

Kim et al. [25] created scaffolding material from PLGA-poly(ethylene glycol) PEG-PLA diblock copolymer from N,N-dimethyl formamide solution and incorporated an antibiotic drug (cefoxitin sodium) which is intended to prevent infection after surgery. The drug has high water-solubility and low solubility in DMF. The release of the drug from the medicated electrospun web into water was monitored, and its effectiveness measured. Attempts were made to slow down the release of the antibiotic because it occurred at a rate that was higher than suitable for medical applications.

Kewany et al. [26] studied drug delivery from PLA, poly(ethylene-*co*-vinyl acetate) (PEVA), and a 1:1 blend of the two polymers electrospun from chloroform solution with tetracycline hydrochloride as the model drug. While PLA delivered the drug more or less instantaneously, the release from PEVA and from the blended nonwoven structures could more appropriately be extended over 5 days. This release rate was much closer to that desired for drug treatment.

A block copolymer of various concentrations of PGA and PEG containing additionally PLGA (75:35, LA:GA) was used as scaffolding material for release

(a)

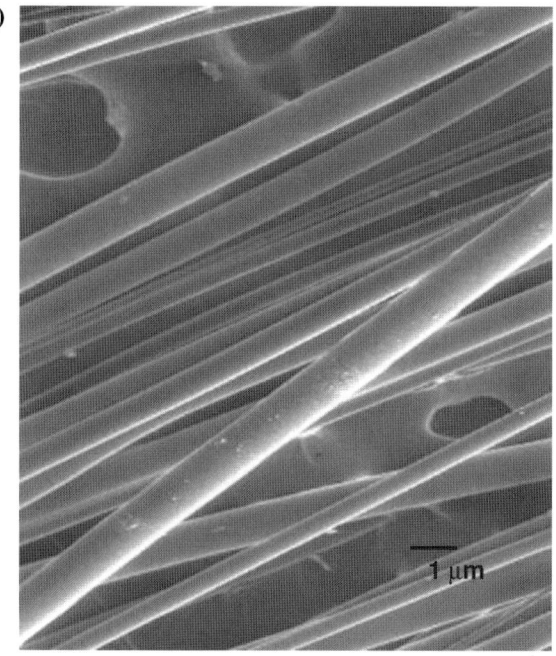

(b) Diameter [μm]

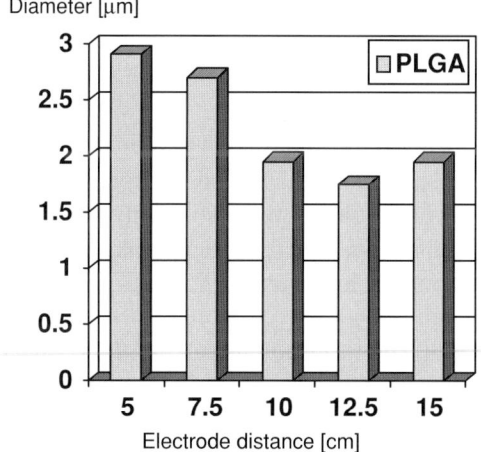

Electrode distance [cm]

Figure 5.3. (a) Electrospun PLGA fibers from melt (15 kV, 7.5 cm electrode distance) and (b) relationship of electrode distance and diameter

of DNA [27]. The mechanical properties of the nonwoven product were tested as well as the structural integrity of the delivered plasmid DNA over 20 days. These results demonstrate that electrospun PLGA is promising as a preliminary candidate for DNA release for gene therapy.

5.4.3 Poly(3-hydroxybutyrate) and copolymers

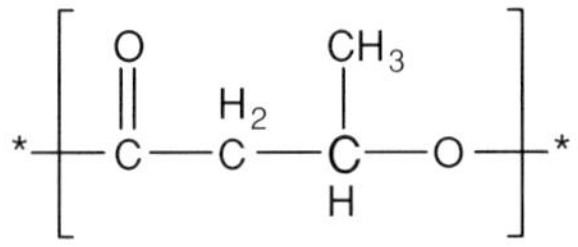

Poly(3-hydroxybutyrate)

Besides PLA, PGA and their copolymers bacterial polyesters poly(3-hydroxybutyrate) PHB and copolymers have been investigated for tissue and cartilage repair as it is compatible with various cell lines [28]. Due to the low mechanical strength of PHB, blends with other polymers have been explored.

PHB can be electrospun from 5% chloroform solution to fibers of less than 1 μm. Figure 5.4 shows fibers produced at a voltage of 10 kV and 7.5 cm electrode distance. The fibers had an average diameter of 860–720 nm (±100), depending on the specific experimental conditions, and a somewhat rough surface.

5.4.4 Poly(ε-caprolactone)

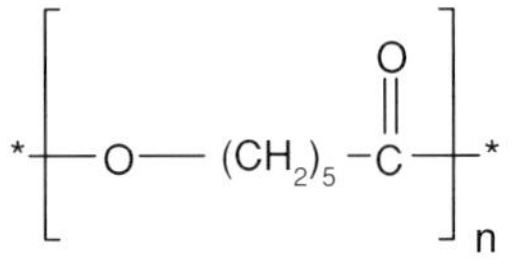

Poly(ε-caprolactone)

Poly(ε-caprolactone) (PCL) has been explored for biomedical applications as well as a potential polymer suitable for electrospinning due to its low melting point of 61 °C and its solubility in various solvents. Thus, PCL can be electrospun from solution as well as from melt. Like PLA, PCL biodegrades in a two-phase process but at a lower rate [29]. It has been recommended for use as a controlled drug delivery system.

Yoshimoto et al. [30] produced electrospun PCL fibers with an average diameter of 400 nm from a chloroform solution with the goal of creating scaffolds for bone tissue engineering. Depending on the spinning conditions the fibers were somewhat irregular in shape along the fiber axis. Their surfaces were also irregular, a characteristic which might aid cell attachment and migration within the scaffolding. Before these fibers were exposed to cell cultures, the PCL fibers were immersed in a collagen solution to encourage cell adhesion. Mineralization of the fibers was achieved within 2 weeks.

5.4.5 Nanoscale silk fibroin fibers

Electrospinning was used to produce silk fibroin fibers from *Bombyx mori* and *Samia cynthia ricini* by dissolving the fibroin in hexafluoroacetone (HFA)

(a)

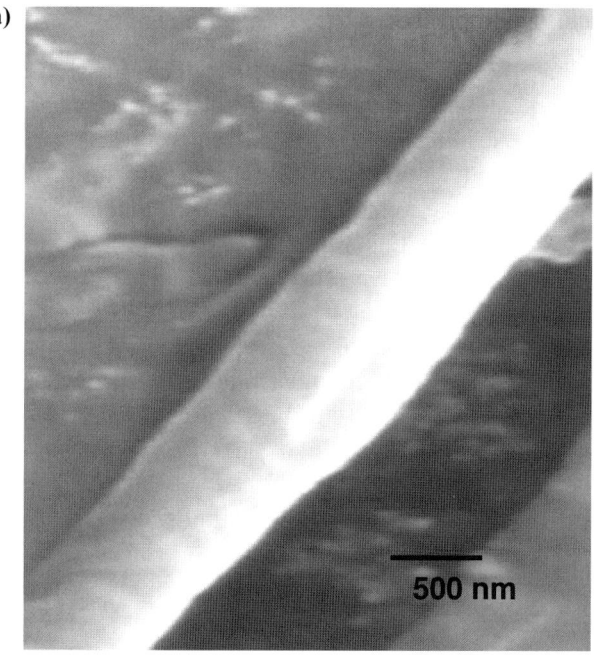

(b) Diameter [μm]

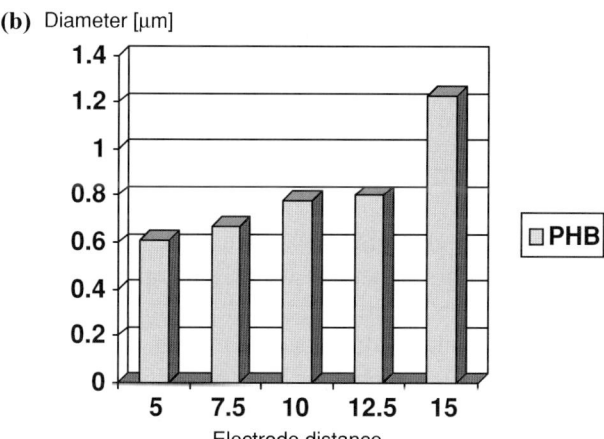

Electrode distance

Figure 5.4. (a) PHB fibers electrospun from 5% chloroform solution and (b) the relationship of diameter and electrode distance

solution [31]. HFA was removed by exposing the electrospun fibers to methanol which served as a coagulant in the post-treatment process. Fiber diameters obtained ranged from 100 to 1000 nm. The research group also reported the production of fibers made from recombinant hybrid silk using a genetic engineering technique. Another possible option to produce nanofibers from silk fibroin by co-spinning with a polymer that electrospins easily was reported by Fridrikh [11].

5.4.6 Cellulose-based nanofibers

Cellulose-based materials are useful for wound dressings and, less importantly, for sutures and related applications. Natural cellulosic materials, such as cotton, decompose before they melt, and cannot be melt-spun. Efforts have been made to regenerate cellulose from solution so as to form nanofibers. Frey [32] successfully produced electrospun cellulose fibers from polar fluid/salt solutions. It is clear from this work that cellulose nanofibers could potentially be spun from inexpensive renewable resources or reclaimed cellulosic material.

Another experimental route to the production of nanofibers based on cellulose consists of derivatizing cellulose and subsequently removing the substituents at the cellulosic hydroxyl groups. For instance, cellulose acetate was electrospun from acetone, acetic acid, and dimethyl acetamide. These solvents and the resulting fibers were studied in connection with the type of collector used [33]. Depending on the composition of the solvent and the concentration of cellulose acetate, fibrous products or beads were obtained. Deacetylation with alcoholic sodium hydroxide solution was performed to various degrees of acetylation from 0.15 to 2.33 without major changes in surface characteristics.

Ding et al. [34] produced blended nanofibrous mats of cellulose acetate and poly(vinyl alcohol) PVA, extruded from separate syringes. Higher mechanical strength was achieved as the content of PVA was increased.

5.5 Conclusions

With the help of the electrostatic spinning technology, biopolymers can be formed into nanofibrous structures which have great potential for medical applications. Due to their small size, the electrospun fibers provide a large surface-to-volume ratio and could be used for drug delivery, scaffolds for tissue engineering, or provide support for bone repair. Due to the relative ease of the electrospinning technique a large number of different polymeric materials including natural fibers have already been explored or will be in the near future. Carefully tailored surface chemistries of these micro- and nanofibers will continue to expand their applications in the medical field.

References

1. Doshi, J.; Reneker, D. H. Electrospinning process and applications of electrospun fibers. *J. Electrostat.* **1995**, *35*(2/3), 151–160.
2. Huang, Z. M.; Zhang, Y. Z.; Kotaki, M.; Ramakrishna, S. A review on polymer nanofibers by electrospinning and their applications in nanocomposites. *Composites Sci. Technol.* **2003**, *63*(15), 2223–2253.

3. Shin, Y. M.; Hohman, M. M.; Brenner, M. P.; Rutledge, G. C. Experimental characterization of electrospinning: the electrically forced jet and instabilities. *Polymer* **2001**, *42*(25), 9955–9967.

4. Formhals, A. Process and apparatus for preparing artificial threads, US Patent 1,975,504, **1934**.

5. Goheen, S. C.; Gaithner, K. A.; Rayburn, A. R. Nature and nanotechnology. In: Tsakalkos, T.; Ovidko, I. A.; Vasuderan, A. K. (Eds.) *Nanostructures: Synthesis, Functional Properties and Applications, Chapt. 3*, National Sci. Series II, Math., Phys., Chem., Vol. 128, wer Publ., Boston, MA, **2003**, 117–137.

6. Deitzel, J. M.; Kleinmeyer, J.; Hirvonen, J. K.; Beck, T. N. C. Controlled deposition of electrospun poly(ethylene oxide) fibers. *Polymer* **2001**, *42*, 8163–8170.

7. Gibson, P.; Schroeder-Gibson, H.; Rivin, D. Transport properties of porous membranes on electrospun nanofibers. *Colloids Surfaces A: Physicochem. Eng. Aspects*, **2001**, 187–188, 469–681.

8. Gibson, P. Production and characterization of patterned electrospun fibrous membranes. Fiber Soc. Meeting Proceedings, **2003**, 42–49.

9. Jaffe, M.; Shanmukasundaram, S.; Patlolla, A.; Griswold, K.; Wang, S.; Mantilla, J.; Walsh, R.; Arinzeh, T.; Prestigiacomo, C. J.; Catalani, L. H. The benefits of nanofibers in biomedical applications, Fiber Soc. Fall Meeting Proceedings, **2004**, 8–9.

10. Park, J. B.; Bronzino, J. D. (Eds.) *Biomaterials, Principles and Applications*. CRC Press, Boca Raton, FL, **2003**.

11. Fridrikh, S. V.; Yu, J. H.; Brenner, M. P.; Rutledge, G. C. Electrostatic production of nanofibers: Physics and applications. Fiber Soc. Fall Meeting Proceedings, **2004**, 37–38.

12. Chung, H. Y.; Hall, J. R. B.; Gogins, M. A.; Crofoot, D. G.; Weik, T. M. Polymer, polymer microfiber, polymer nanofiber and applications including filter structures. PCT Int. Appl., **2002**, WO 2001-US24948 20010809.

13. Reneker, D. H.; Chun, I. Nanometer diameter fibres of polymer, produced by electrospinning. *Nanotechnology* **1996**, *7*(3), 216–223.

14. Lee, S. H.; Yoon, J. W.; Moon, H. Continuous nanofibers manufactured by electrospinning technique. *Macromol. Res.* **2002**, *10*(5), 282–285.

15. Chu, C. C.; von Fraunhofer, J. A.; Greisler, H. P. (Eds.) *Wound Closure Materials and Devices*. CRC Press LLC, Boca Raton, FL, **1997**.

16. Dee, K. C.; Puleo, D. A.; Bizios, R. (Eds.) *An Introduction to Tissue–Biomaterial Interactions*, Chapt. 2. Wiley-Liss, Hoboken, NJ, **2002**.

17. Brodsky, B.; Ramshaw, J. A. M. The collagen triple-helix structure. *Matrix Biol.* **1997**, *15*, 545–554.

18. Matthews, J. A.; Wnek, G. E.; Simpson, D. G.; Bowlin, G. L. Electrospinning of collagen nanofibers. *Biomacromolecules* **2002**, *3*, 232–238.

19. Simpson, D. G.; Bowlin, G. L.; Wnek, G. E.; Stevens, P. J.; Carr, M. E.; Matthews, J. A.; Rajendran, S. Electroprocessed collagen and tissue engineering, US patent application 0040037813, **2004**.

20. Hawkins, A.; Wood, J.; Buschle-Diller, G. Advances in electrospinning of biopolymers. AATCC Intern. Convention & Exhibition, Greenville, SC, Sept. 13–17, **2004**.

21. Kidoaki, S.; Kwon, I. K.; Mattsuda, T. Mesoscopic spatial design of nano- and microfiber meshes for tissue-engineering matrix and scaffold based on newly devised multilayering and mixing electrospinning techniques. *Biomaterials* **2004**, *26*, 37–46.

22. Boland, E. D.; Wnek, G. E.; Simpson, D. G.; Pawlowski, K. J.; Bowlin, G. L. Tailoring tissue engineering scaffolds by employing electrostatic processing techniques: A study of poly(glycolic acid). *J. Macromol. Sci.* **2001**, *38*, 1231–1243.

23. Bognitzki, M.; Czado, W.; Frese, T.; Schaper, A.; Hellwig, M.; Steinhardt, M.; Greiner, A.; Wendorff, J. H. Nanostructured fibers via electrospinning. *Adv. Mater.* **2001**, *13*(1), 70–72.

24. Zong, X.; Kim, K.; Fang, D.; Ran, S.; Hsaio, B. S.; Chu, B. Structure and process relationship of electrospun bioabsorbable nanofiber membranes. *Polymer* **2002**, *43*(16), 4403–4412.

25. Kim, K.; Luu Y. K.; Chang, C.; Fang, D.; Hsiao, B. S.; Chu, B.; Hadjiargyrou, M. Incorporation and controlled release of a hydrophilic antibiotic using poly (lactide-*co*-glycolide)-based electrospun nanofibrous scaffolds. *J. Control. Release* **2004**, *98*(1), 47–56.

26. Kewany, E. R.; Bowlin, G. L.; Mansfield, K.; Layman, J.; Simpson, D. G.; Sanders, E. H.; Wnek, G. E. Release of tetracycline hydrochloride from electrospun poly (ethylene-*co*-vinyl acetate), poly (lactic acid) and a blend. *J. Control. Release* **2002**, *81*(1–2), 57–64.

27. Luu, Y. K.; Kim, K.; Hsioa, B. S.; Chu, B.; Hadjiargyrou, M. Development of a nanostructured DNA delivery scaffold via electrospinning of PLGA and PLA-PEG block copolymers. *J. Control. Release* **2003**, *89*(2), 341–353.

28. Zhao, K.; Deng, Y.; Chen, J. C.; Chen, G. Q. Polyhydroxyalkanoate (PHA) scaffolds with good mechanical properties and biocompatibility. *Biomaterials* **2003**, *24*(6), 1041–1045.

29. Cha, Y.; Pitt, C. G. The biodegradability of polyester blends. *Biomaterials* **1990**, *11*(2), 108–112.

30. Yoshimoto, H.; Shin, Y. M.; Terai, H.; Vacanti, J. P. A biodegradable nanofiber scaffold by electrospinning and its potential for bone tissue engineering. *Biomaterials* **2003**, *24*, 2077–2082.

31. Ohgo, K.; Zhao, C.; Kobayashi, M.; Asakura, T. Preparation of non-woven nanofibers for *Bombyx mori* silk, *Samia cythia ricini* silk and recombinant hydrid silk with electrospinning method. *Polymer* **2003**, *44*(3), 841–846.

32. Frey, M. W.; Joo, Y. L.; Kim, C.-W. New solvents for cellulose electrospinning and preliminary nanofiber spinning results. In: Symposium on Polymeric Nanofibers, ACS Fall Meeting (**2003**), New York, New York.

33. Liu, H.; Hsieh, Y. L. Ultrafine fibrous cellulose membranes from electrospinning of cellulose acetate. *J. Polym. Sci.: Part B: Polym. Phys.* **2002**, *40*, 2119–2129.

34. Ding, B.; Kimura, E.; Sato, T.; Fujita, S.; Shiratori, S. Fabrication of blend biodegradable nanofibrous nonwoven mats via multi-jet electrospinning. *Polymer* **2004**, *45*(6), 1895–1902.

Chapter 6

HALAMINE CHEMISTRY AND ITS APPLICATIONS IN BIOCIDAL TEXTILES AND POLYMERS

Gang Sun[1] and S. D. Worley[2]

[1]*Division of Textiles and Clothing, University of California, Davis, CA 95616, U.S.A.*
[2]*Department of Chemistry, Auburn University, Auburn, AL 36849, U.S.A.*

6.1 Introduction

Healthcare and emergency workers urgently need personal bioprotective equipment to prevent infections of pathogens such as SARS and biological agents in working environments. Such a need has stimulated research in antimicrobial textiles and polymers, and has led to development of bioprotective clothing materials and polymers that can offer instant biocidal functions against all major pathogens in recent years [1–6]. Biocidal functions can be divided into sterilization, disinfection, and sanitization in an order of the strength of the function. Biocidal functions for personal protection should be at least at the disinfection level, which can inactivate most infectious microorganisms. On the other hand, biocidal clothing materials should be safe to wear and environmentally friendly. Among many available biocides, N-halamine compounds possess superior disinfection power and safety; in fact, many of them are used as swimming pool disinfectants [7]. In addition, N-halamines have demonstrated the capability of providing rapid and total inactivation of a wide range of microorganisms without causing the microorganisms to develop resistance to them [2].

N-Halamine chemistry can be expressed in Eqs. (1) and (2). When N-halamine structures are exposed to water, the reaction shown in Eq. (1) may occur. The equilibrium in Eq. (1) may shift toward either reactants or products depending on the N-halamine structures. Three principle N-halamine

81

J. V. Edwards et al. (eds.), Modified Fibers with Medical and Specialty Applications, 81–89.

structures: imide, amide, and amine, are available, and their stability and dissociation constants in water are shown

$$\ce{>N-Cl + H2O <=> >N-H + Cl+ + OH-} \qquad (1)$$

$$\ce{>N-Cl <=>[Kill bacteria][Bleach] >N-H} \qquad (2)$$

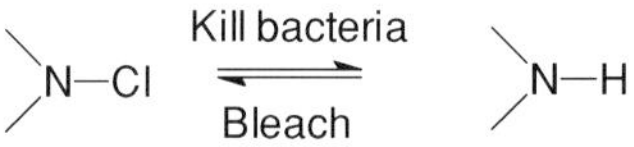

in Table 6.1 [4]. *N*-Halamine structures can kill microorganisms directly also without the release of free chlorine, as in Eq. (2) [8]. In fact, according to their dissociation constants shown in Table 6.1, *N*-halamine structures such as the amines may only release negligible amounts of free chlorine. Since *N*-halamine structures are biocidal, and more importantly quite stable in ambient environments, incorporation of the *N*-halamine into polymeric and textile materials will bring biocidal functions to them. Moreover, since Eq. (2) is a reversible reaction, the biocidal functions on the materials are rechargeable with a chlorinating agent, such as chlorine bleach. This rechargeable function is most suitable for reusable medical textiles and clothing. In this chapter, we will review the latest progresses in the application of *N*-halamine chemistry to textiles and polymers.

Table 6.1. Stability of *N*-halamine structures [4]. (Journal of Applied Polymer Science © 2003)

Dissociation reaction	Dissociation constant for examples
Imide Structure 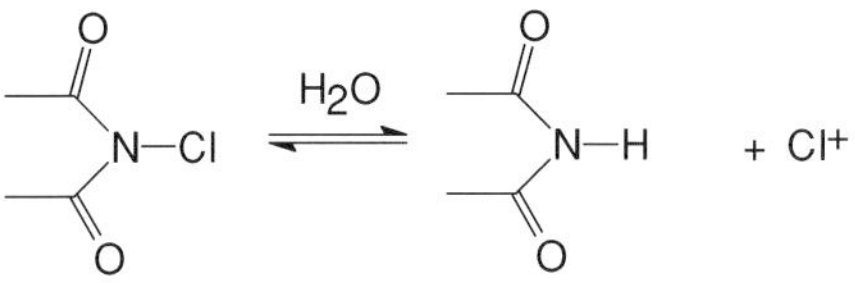	$1.6 \times 10^{-2} - 8.5 \times 10^{-4}$, Trichlorocyanuric acid 2.54×10^{-4}, 1,3-dichloro-5, 5-dimethylhydantoin
Amide Structure 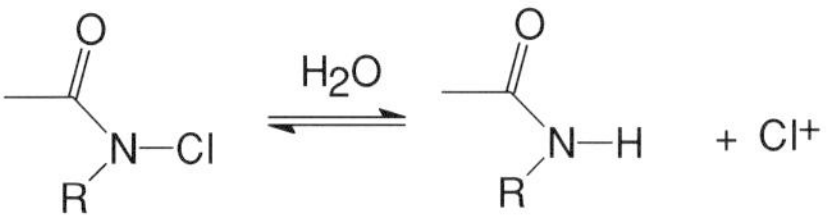	2.6×10^{-8}, 1,3-dichloro-2,2,5, 5-tetramethyl-4-imidazolidinone 2.3×10^{-9}, 3-chloro-4,4-dimethyl-2-oxazolidinone
Amine Structure 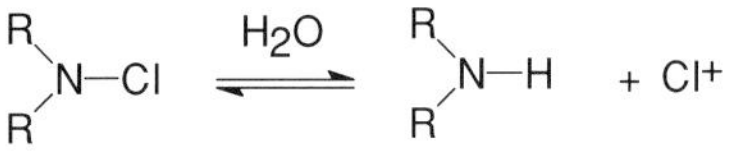	$< 10^{-12}$

1,3,-dimethylol-5,5-dimethylhydantoin   3-methylol-2,2,5,5-tetramethylimidazolidin-4-one
(DMDMH)   (MTMIO)

Figure 6.1. Structures of DMDMH and MTMIO

6.2 Incorporation of *N*-halamine in cellulose

6.2.1 DMDMH-treated cellulose

Both amide and imide *N*-halamines have been incorporated into cellulose-containing fabrics by a conventional finishing method with 1,3-dimethylol-5,5-dimethylhydantoin (DMDMH, shown in Figure 6.1) [3]. Although it would appear from the structure in Figure 6.1 of DMDMH that there is no empty nitrogen site for oxidative chlorine to bind, in practice some loss of formaldehyde occurs during the treatment process providing sites for chlorination with chlorine bleach. The DMDMH-treated fabrics exhibited rapid biocidal functions, but the washing durability of the functions requires improvement, due to the dominating imide *N*-halamine functionality, which is the most reactive, but least stable on the fabrics. However, DMDMH fabrics can be employed in personal protection against various biological agents such as bacteria, viruses, fungi, yeasts, and spores [3, 9]. Examples of the treated fabrics demonstrated a complete elimination of pathogens in a contact time as short as 2 min [2–3]. The biocidal functions could be recharged repeatedly for at least 50 machine washes.

6.2.2 MTMIO-treated cellulose

In order to increase washing durability of the *N*-halamine-treated textiles, the more stable amine *N*-halamine has been grafted to cellulose in a similar approach by using 3-methylol-2,2,5,5-tetramethylimidazolidin-4-one (MTMIO, shown in Figure 6.1) [4]. The resulting fabrics contained the more stable, and less reactive, amine *N*-halamine structure, thus providing slow, but durable, biocidal functions. The improved biocidal functions are summarized in Table 6.2 which compares fabrics treated by DMDMH and MTMIO separately. Both active chlorine contents and biocidal functions against *Escherichia coli* (*E. coli*) and *Staphylococcus aureus* (*S. aureus*) are listed in the table.

Table 6.2. Chlorine loss and anti-microbial effects of MTMIO- and DMDMH-modified cotton samples

Chemical	Washing cycles	Against *E. coli*			Against *S. aureus*		
		Cl (ppm)	Cl loss (%)	Log reduction	Cl (ppm)	Cl loss (%)	Log reduction
	0	565	—	6	654	—	6
MTMIO	2	507	10.2	5	616	6.1	6
	5	498	11.9	4	601	8.4	4
	0	863	—	6	934	—	6
DMDMH	2	218	74.7	1.5	380	59.3	3
	5	157	81	0.9	274	70.7	2

Pure cotton fabric 493#; total finishing bath concentration: 4%. Wet pick-up: 70%. Concentrations of bacteria: *E. coli* 5×10^6 CFU/mL and *S. aureus* 7×10^6 CFU/mL. A six log reduction is equivalent to 99.9999% inactivation. Contact time: 60 min. Machine washing according to AATCC standard test method 124-1999; tests 1 and 2. The MTMIO-treated fabric was bleached separately from the DMDMH-treated fabric with the same concentration of active chlorine (150 ppm) used in each case (Journal of Applied Polymer Science © 2003).

6.2.3 Hydantoinylsiloxane-treated cellulose

Biocidal functionality can also be introduced into cellulose by condensing the hydroxyl groups on 3-trihydroxysilylpropyl-5,5-dimethylhydantoin (SPH) (Figure 6.2) with those on cellulose followed by chlorination of the amide nitrogen on the hydantoin ring with chlorine bleach [10]. It was observed that a complete 5.7 log reduction of *S. aureus* could be obtained in a contact-time interval of 30–60 min. Likewise, a complete 5.9 log reduction of *E. coli* was observed in a contact-time interval of 60–120 min. The chlorine loading on the cotton cloth was 0.5–0.6% by weight in these experiments. For comparison purposes, cotton cloth was also treated with the quaternary ammonium compound dimethyloctadecyltrimethoxysilylpropylammonium chloride. In this case the log reductions of *S. aureus* and *E. coli* were only 1.8 and 2.5, respectively, in

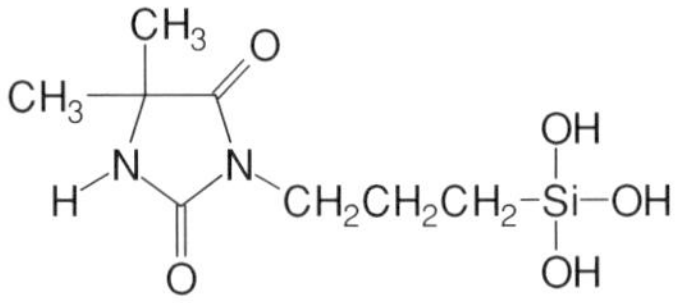

3-trihydroxysilylpropyl-5,5-dimethylhydantoin
(SPH)

Figure 6.2. Structure of SPH

the same contact-time intervals as those tested for the samples treated with SPH. This comparison demonstrates conclusively the superiority of cellulose treated with *N*-halamines over that treated with biocidal quaternary ammonium salts. The chlorinated SPH-treated cloth is reasonably stable to loss of chlorine during dry storage. A loss from 0.62% to 0.54% Cl was observed over a 50-day period for the treated cloth stored in a non-airtight plastic bag. In standard washing tests it was found that cotton cloth treated with SPH and chlorinated with an initial chlorine loading of 0.61% retained 0.42% Cl after 5 machine washings, 0.41% after 10 washings, and 0.10% after 50 washings; thus the material still retained some biocidal functionality even after 50 machine washings.

The SPH compound has also been employed to treat commercial office envelope paper [10]. In this case, a chlorine loading of 0.82% by weight was obtained. The chlorine content declined only to 0.78% over a 36-day period of storage in a vacuum desiccator. The paper completely inactivated *S. aureus* (5.4 logs) at a contact time of only 10 min.

6.3 Incorporation of *N*-halamines in other textile materials

6.3.1 ADMH-treated fibers

Recently, a hydantoin-containing monomer, 3-allyl-5,5-dimethylhydantoin (ADMH, as shown in Scheme 6.1) was prepared to incorporate only amide *N*-halamine structures into synthetic fibers [11, 12]. Due to the amide structure, the thus-produced fabrics could demonstrate both powerful and durable biocidal functions. Synthetic fabrics such as nylon-66, polyester (PET), polypropylene (PP), acrylics, Nomex IIIa, PBI/Kevlar, and Kermel, as well as pure cotton fabrics, were used in the chemical modification. The ADMH can be incorporated in surfaces of fibers by a controlled radical grafting reaction which can ensure short chain grafts instead of long chain self-polymerization of the monomers.

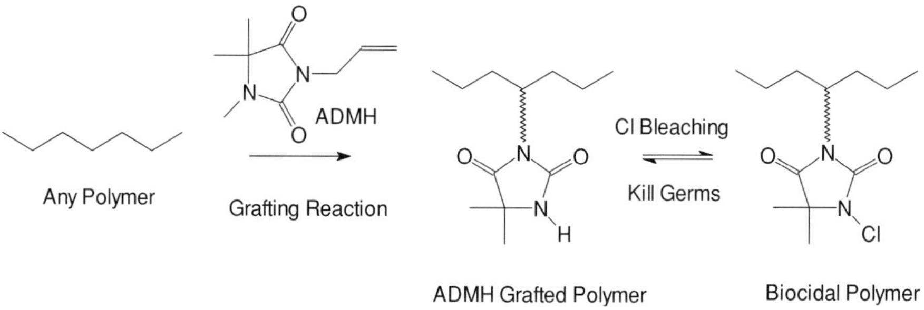

Scheme 6.1. Structure of ADMH and its grafting reactions on synthetic polymers

6.3.2 Grafting polymerization

Both water-soluble and water-insoluble initiators such as potassium persulfate (PPS) and benzoyl peroxide (BPO) were used to initiate radical grafting reactions on fibers. Water-soluble initiators can specifically work on hydrophilic fibers such as cellulose, while water-insoluble BPO was quite effective on hydrophobic fibers such as polyester and most synthetic fibers. The fabric treatment was conducted in a convenient wet finishing process involving pad-dry-cure. Fabrics were immersed in the chemical solutions and padded under pressure to expel additional liquid. The fabrics were dried at 50 °C for 5 min, cured at an elevated temperature for a specific period of time, and then washed with a large amount of distilled water, dried at 60 °C for 24 h, and stored in a conditioned room (25 °C, 65% RH) for 48 h to reach constant weight. The chemical reactions involved in this process are shown in Scheme 6.1 [11, 12].

6.3.3 Observed results

Biocidal properties of the modified fibers could be demonstrated after a chlorination reaction by exposing the grafted fibers to a diluted chlorine solution, with which the grafted hydantoin rings were converted to N-halamine structures. The polymeric N-halamines could provide powerful and rapid antibacterial activities against *E. coli* and *S. aureus*. Most of the fibers could completely inactivate a large number of bacteria (1×10^6 CFU) in a 10–30 min contact time. In addition, the anti-bacterial activities of these polymeric N-halamines could be easily recovered after usage by simply exposing to chlorine solution again.

Figure 6.3 reveals grafting yields on common fabrics using BPO as radical initiator and triallyl-1,3,5-triazine-2,4,6(1H,3H,5H)-trione (TATAT) as a co-monomer in the grafting reactions [12]. These fabrics represent the most commonly employed fabric materials in institutional and consumer uses. The grafting reactions were almost quantitative for several fabrics. However, since BPO is very hydrophobic, it did not work as effectively on cotton fabrics as did hydrophilic radical initiators [12]. Table 6.3 shows the durability of the biocidal functions provided by the ADMH grafted fabrics. The biocidal functions on most hydrophobic fabrics were more durable than those on hydrophilic fabrics and could last for more than 15 washes without recharging. After 50 washes, the lost biocidal functions were fully recharged by a dilute chlorine washing.

The controlled radical grafting reaction also worked effectively on some high performance fabrics such as Nomex, Kermel, and Kevlar/PBI, which are often employed in firefighter and military uniforms. Shown in Figure 6.4 are

Table 6.3. Log reduction of *E. coli* after washing. (Journal of Applied Polymer Science © 2002)

Washing times	Log reduction of *E. coli* (%)					
	Nylon	PET	PP	Acrylic	Cotton	PET/Cotton
0	5	5	5	5	5	5
5	5	5	5	5	3	5
15	5	5	5	5	1	5
30	3	3	2	1	UD[a]	3
50	UD	1	1	1	UD	UD
50[b]	5	5	5	5	5	5

Note: contact time = 30 min (*E. coli* concentration: 10^5–10^6 CFU/mL; all of the samples were tested with machine washing following AATCC Test Method 124. AATCC standard reference detergent 124 was used in all of the machine-washing tests.
[a] No reduction of *E. coli* was detected.
[b] These samples were re-bleached after 50 times of washing.

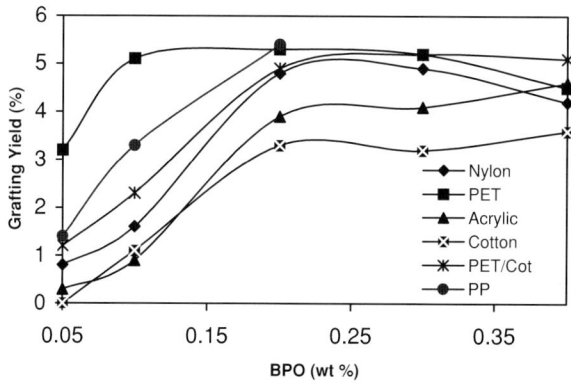

Figure 6.3. Influence of BPO concentration (wt%) on grafting yields [12]. (Journal of Applied Polymer Science © 2002) Grafting conditions: Padding bath contained ADMH, 4 wt%; TATAT, 1.5 wt%; the softener, 1.5 wt%; and different amounts of BPO. The fabrics were dipped and padded twice at a 100% expression, dried at 50 °C for 5 min, cured at 130 °C for 5 min (for PP, the fabric was cured at 105 °C for 5 min), washed, and dried

grafting yields of ADMH on these fabrics. The reaction on Nomex was especially highly efficient with the grafting yields above 4%. This result may be caused by lower crystallinity of the polymer than the other two since the reaction could only occur in amorphous areas of the polymers. Since these fabrics are quite hydrophobic, the adsorption of ADMH on the polymers was relatively difficult, which may contribute to overall lower grafting yields and low biocidal functions on Kermel and Kevlar/PBI fabrics.

Hydrophobic properties may prevent hydrolysis of *N*-halamine structures on the fabrics. Thus, these fabrics all demonstrated quite durable biocidal

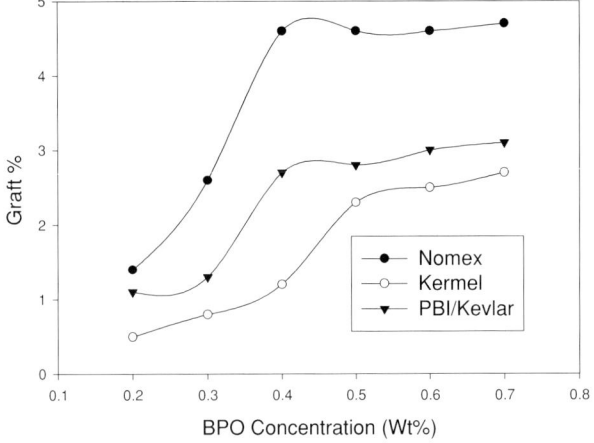

Figure 6.4. Grafting yields on Nomex, Kermel, and Kevlar/PBI fabrics. (Journal of Applied Polymer Science © 2003) Padding bath contained: ADMH, 3 wt%; PEG-DIA, 2 wt%; and the softener, 1.5 wt%. The fabrics were dipped and padded twice at a 100% expression, dried at 50 °C for 5 min, cured at 140 °C for 5 min, washed, and dried.

functions, as shown in Table 6.4. The grafting reactions on Kermel and Kevlar/PBI fabrics were less effective, as observed by the low-active chlorine contents on the fabrics, and low biocidal efficacy.

6.3.4 Condensation reactions

It has also been demonstrated that nylon-66 and PET can be rendered biocidal by utilization of N-halamine chemistry analogous to that discussed for

Table 6.4. Log reduction of the bacteria after washing at a contact time of 60 min (bacteria concentration: 10^6–10^7 CFU/mL). (Journal of Applied Polymer Science © 2003)

Wash times	Nomex® $M_{Cl} \times 10^5$ (mol/g)	*E. coli*	*S. aureus*	Kermel® $M_{Cl} \times 10^5$ (mol/g)	*E. coli*	*S. aureus*	PBI®/Kevlar® $M_{Cl} \times 10^5$ (mol/g)	*E. coli*	*S. aureus*
0	1.22	6	6	0.33	3	5	0.41	3	6
5	1.20	6	6	0.28	3	4	0.41	3	4
15	0.63	6	5	0.23	3	2	0.37	3	2
30	0.27	3	4	UD[a]	1	1	0.20	1	2
50	UD[a]	1	1	UD[a]	UD[a]	UD[a]	UD[a]	UD[a]	UD[a]
50[b]	1.14	6	6	0.29	3	5	0.43	3	6

[a]No reduction was detected.
[b]These samples were re-bleached after 50 times of washing.

DMDMH in section 6.2.1 [13, 14] and the hydantoinylsiloxane in section 6.2.3 [10].

Acknowledgments

The research was support by a CAREER award from The National Science Foundation (DMI 9733981) (GS), The National Textile Center (C02-CD06) (GS), The U.S. Air Force (FO8637-02-C-7020) (SDW), and Vanson-HaloSource Inc. (GS and SDW).

References

1. Worley, S. D.; Sun, G. Biocidal polymers. *Trends Polym. Sci.* **1996**, *4*, 364–370.
2. Sun, G.; Xu, X. Durable and regenerable antibacterial finishing of fabrics. Biocidal properties. *Text. Chem. Colorist* **1998**, *30*, 26–30.
3. Sun, G.; Xu, X.; Bickett, J. R.; Williams, J. F. Durable and regenerable antimicrobial finishing of fabrics with a new hydantoin derivative. *Ind. Eng. Chem. Res.* **2001**, *41*, 1016–1021.
4. Qian, L.; Sun, G. Durable and regenerable antimicrobial textiles: Synthesis and applications of 3-methylol-2,2,5,5-tetramethyl-imidazolidin-4-one (MTMIO). *J. Appl. Polym. Sci.* **2003**, *89*, 2418–2425.
5. Sun, Y. Y.; Chen, T. Y.; Worley, S. D.; Sun, G. Novel refreshable *N*-halamine polymeric biocides containing imidazolidin-4-one derivatives. *J. Polym. Sci. Part A: Polym. Chem.* **2001**, *39*, 3073–3084.
6. Chen, Y.; Worley, S. D.; Kim, J.; Wei, C.-I.; Chen, T. Y.; Santiago, J. I.; Williams, J. F.; Sun, G. Biocidal poly(styrenehydantoin) beads for disinfection of water. *Ind. Eng. Chem. Res.* **2003**, *42*, 280–284.
7. Worley, S. D.; Williams, D. E. Halamine water disinfectants. *CRC Crit. Rev. Environ. Control* **1988**, *18*, 133.
8. Williams, D. E.; Elder, E. D.; Worley, S. D. Is free halogen necessary for disinfection? *Appl. Environ. Microbiol.* **1988**, *54*, 2583–2585.
9. Weber, D. J.; Sickbert-Bennett, E.; Gergen, M. F.; Rutala, W. A. Efficacy of selected hand hygiene agents used to remove *Bacillus atrophaeus* (a surrogate of *Bacillus anthracis*) from contaminated hands. *J. Am. Med. Assoc.* **2003**, *289*, 1274–1277.
10. Chen, Y.; Worley, S. D. Unpublished data. **2003**.
11. Sun, Y. Y.; Sun, G. Durable and regenerable antimicrobial textile materials prepared by a continuous grafting process. *J. Appl. Polym. Sci.* **2002**, *84*, 1592–1599.
12. Sun, Y. Y.; Sun, G. Novel refreshable *N*-halamine polymeric biocides: Grafting hydantoin-containing monomers onto high-performance fibers by a continuous process. *J. Appl. Polym. Sci.* **2003**, *88*, 1032–1039.
13. Lin, J.; Winkelmann, C.; Worley, S. D.; Broughton, R. M.; Williams, J. F. Antimicrobial treatment of nylon. *J. Appl. Polym. Sci.* **2001**, *81*, 943–947.
14. Lin, J.; Winkelmann, C.; Worley, S. D.; Kim, J.; Wei, C-I.; Cho, U.; Broughton, R. M.; Santiago, J. I.; Williams, J. F. Biocidal polyester. *J. Appl. Polym. Sci.* **2002**, *85*, 177–182.

Chapter 7

MODIFICATION OF POLYESTER FOR MEDICAL USES

Martin Bide[1], Matthew Phaneuf[2], Philip Brown[3], Geraldine McGonigle[4], and Frank LoGerfo[5]

[1] University of Rhode Island, Kingston, RI, U.S.A.
[2] BioSurfaces, Ashland, MA, U.S.A.
[3] Clemson University, Clemson, SC, U.S.A.
[4] Unilever R&D, Port Sunlight Laboratories, Bebington, UK
[5] Beth Israel Deaconess Medical Center, Boston, MA, U.S.A.

7.1 Introduction/Overview

Polyester has been around for more than half a century, in which time its cost and properties have combined to make it the world's best selling synthetic fiber. As with any other manufactured fiber, it is routinely sold in many variants of diameter, cross-section, and luster. It has a number of limitations, but its ubiquity has lead to many attempts to modify the fiber to overcome them.

Among its many uses are several in the medical field: polyester is a very useful biomaterial. In routine use, however, and similar efforts to modify the fiber have been widely researched.

This chapter reviews the modifications applied to polyester in normal use, the principles by which they have their effects, and the parallel modifications used for medical applications of the fiber.

7.2 Polyester

Poly(ethylene terephthalate), PET, or most simply "polyester" was first discovered in 1941 in the Accrington, UK laboratories of The Calico Printers Association. Whinfield and Dickson demonstrated that a partially aromatic polyester based on terephthalic acid yielded a strong, resistant fiber with a high melting point and good hydrolytic stability [1]. The fiber was commercialized

91

J. V. Edwards et al. (eds.), Modified Fibers with Medical and Specialty Applications, 91–124.

through the 1950s as Dacron™ in the United States and as Terylene™ in the United Kingdom. It entered a textile world in which nylon had been around for a decade or more, and nylon had pioneered both the replacement of natural fibers in traditional end uses and the development of novel synthetic fiber uses. The production of PET accelerated in the early 1960s when processes to produce pure terephthalic acid were improved [2]. The 1970s saw a fashion-led "polyester boom" that made consumers aware of both the good and bad properties of polyester: the latter tended to predominate and polyester acquired connotations of cheapness and tastelessness that have only recently been overcome. Meanwhile, with a growing volume of production, the lower cost of polyester allowed it to replace other synthetic fibers, especially nylon and acrylic, in many routine textile applications where these did not have an advantage in properties. Its esthetic qualities have been improved largely by the development of finer fibers, especially the so-called "microfibers", and today polyester is approaching cotton as the world's most widely used fiber: more than 24 million tons of polyester were produced in 2004.

7.2.1 Production and properties

PET is a linear macromolecular homopolymer (i.e., one repeating unit) formed from step reaction polymerization. It is nominally produced by the polymerization of terephthalic acid and ethylene glycol. These monomers are both readily available in few reaction steps from the petroleum-based feedstocks of the refinery: terephthalic acid from the oxidation of *p*-xylene, ethylene glycol from ethylene *via* ethylene oxide, and the low monomer cost contributes to the low cost of the final polymer.

In practice, terephthalic acid is converted first either to dimethyl terephthalate, or *bis*(hydroxyethyl) terephthalate, and the polymerization proceeds with the elimination (condensation) of either methanol or excess ethylene glycol as the ester interchange reaction continues. PET is predominantly produced using antimony III catalysts, about which there are on-going environmental concerns [3]. Titanium IV catalysts have superior activity to antimony III, but for many years the polymers produced using this technology have had a limited attainable molecular weight and also exhibited a yellowish tinge. However, the thrust for research into catalyst development is not founded on environmental concerns alone. The desire to increase plant capacity and improve product quality and economy by reducing the additive levels is an alternate goal. For a new catalyst technology platform to be commercially acceptable, it must possess certain basic qualities. These include, but are not limited to, good solubility in ethylene glycol as well as stability to water, phosphorus stabilizers, and other necessary additives. There are problems associated with titanium catalysts, which have to be solved, for example, the well documented tendency for hydrolysis that destroys the catalyst, causing thermal instability and haze in the resultant

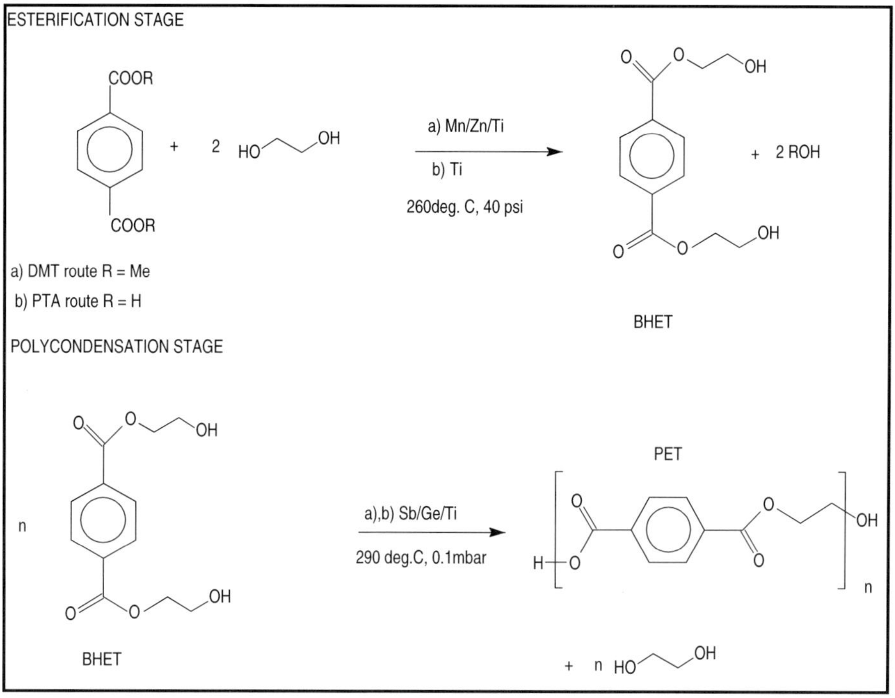

Figure 7.1. Production of polyester

polymer. The polymer should exhibit equivalent color, clarity, and stability to that of standard PET. The basic chemical pathways for PET production are shown in Figure 7.1.

The molecular weight of PET is most often described in the PET industry in terms of intrinsic viscosity (IV) in units of dL/g measured by solution viscosity in a suitable solvent such as *o*-chlorophenol at 25 °C. The range of IVs typically used depends upon the end use of the PET fibers. For example for apparel applications, an IV of 0.63 dL/g is typical but for industrial yarn applications, higher IV polymers are used in order to obtain higher tenacities. Depending upon the industrial end use, the IV range used to make these yarns will fall between 0.80 and 0.92 dL/g.

Fiber is produced by the extrusion of a polymer melt through spinnerets. The absence of solvents in the fiber spinning process is a further contributor to the low cost of the fiber. Polyester can be produced in a wide range of deniers; finer ones (below about 0.5 denier) require modified spinning methods. The fiber cross-section can be modified by changing the shape of the spinneret holes; and while a round section is usual, a myriad of other shapes have been produced, the more common of which would be trilobal or pentalobal. Most fiber is rendered semi-dull or dull by the inclusion of titanium dioxide delustrant. The fiber is solidified by cooling and drawn to encourage polymer chain orientation and crystallinity. The fiber is produced as continuous filaments that can be texturized, or cut into staple lengths. Filaments may be partially drawn (POY) for later draw texturizing, while FOY is fully drawn at the fiber production stage.

Polyester's general properties are well established [4]. Polyester melts at around 260 °C. Its glass-transition temperature of around 100°C is high compared with other thermoplastic fibers, and this limits the action of many reagents to the fiber surface at temperatures below T_g. This is particularly apparent when the fiber is dyed: the wide use of polyester prompted the development and implementation of dyeing machines capable of dyeing at elevated pressures (and thus temperatures): suitable dyes readily penetrate the fiber at 130°C. Depending on the degree of drawing, polyester has a tenacity of 4–8 g/d and an elongation of 40–20% (higher tenacities are associated with lower elongations). The standard moisture regain is around 0.5%, and the low sorption of moisture means that elongation and tenacity vary little when the fiber is wet or dry. Fabrics of polyester have good abrasion resistance and good recovery and resilience.

Polyester's all-round good properties, and its low cost, have made it ubiquitous. It is used in all fabric styles from sheer georgette to heavy fleece, from stretch knit to canvas. It is used in all types of apparel, in home furnishings, in geotextiles, industrial applications, tire reinforcement, and so on. Somewhat like cotton among the natural fibers, the first choice for a synthetic fiber would probably be polyester, and only if that does not fit the bill would another fiber be considered.

7.3 Medical uses of polyester

The biological response of a wide range of (mostly fibrous) materials was studied in the early 1950s, and polyester's superiority was apparent. Good strength, the lack of a reaction by the body, and the inaccessibility of the fiber interior to potentially degrading materials make polyester a useful and biodurable material. There are over 13 million medical devices implanted annually in the United States, ranging from simple devices such as hernia repair mesh, wound dressings, and catheter cuffs to more complex devices such as the total implantable heart, left ventricular assist devices, and prosthetic arterial grafts [5]. When textile fibers must be left in the body for extended times, polyester is useful. In vascular grafts, polyester has been used in woven, warp, and weft-knitted structures, both smooth and texturized: in all these, the major variation is in fabric, rather than fiber structure. The initial development of vascular grafts, their structure, and performance up to the early 1980s was reviewed [6]. Structurally, little further advance has taken place since then, but the need for improvement has prompted much work on modifying the polyester from which they are constructed, as noted in what follows. An extensive review of arterial grafts, their limitations, and modifications has been published [7].

Prosthetic heart valves require a textile-based cuff that can be used to sew the valve into place. The experience with vascular grafts suggested that polyester would be an appropriate material in this use, and the suggestion has been borne out in practice. Sewing cuffs are typically knitted structures.

A hernia is an abnormal bulging of internal organs, often the intestine, through a muscular wall weakness. This complication, which can present in several areas of the body such as the stomach, groin, or throat, requires that the inner contents of the muscular wall (e.g., organs) be placed back into the cavity and the tissue repaired *via* suturing. Hernia repair mesh can be made of polyester, PTFE, or polypropylene and be designed into various shapes. This is sewn into place over the defect to add strength to the repaired tissue or can serve as a tissue substitute in the event of significant structural loss. These devices, which are not exposed to flowing blood as described for the sewing cuff, are designed to last long-term.

As with non-medical uses, other fibers (silk, nylon) are preferred where stretch/recovery is important, and these are still preferred for sutures where long-term durability is not an issue.

7.4 Limitations of polyester in routine (non-medical) use

Polyester has inherent characteristics that limit its use in some applications. Abrasion resistance, recovery, and resilience are generally better in nylon, and nylon retains its wider use in, for example, carpets, hosiery, and ropes for dynamic applications. Its hydrophobic character limits the comfort of polyester garments in warm weather and the absorption characteristics of hydrophilic fibers such as cotton mean that cotton still dominates the apparel and domestic (sheets/towels) textile market. The hydrophobicity of polyester goes hand in hand with oleophilicity, and when developed into a fabric, polyester textiles dry quickly but tend to retain oily soils once again favoring more hydrophilic materials. Polyester fabrics made from staple fibers can develop surface pills: the strength of the fiber tends to hold such pills on the fabric when a weaker fiber would allow them to break away.

7.5 Modifications of polyester in routine (non-medical) use

In routine use, polyester can be modified for a number of reasons and by several different strategies. The principles involved and the techniques based on those principles have, in many cases, given rise to parallel modifications in the medical uses of polyester. An extensive survey of such modifications is provided: if they have not yet been used to modify medical polyester, they may suggest future opportunities. Progress up to around 1980 in routine use modifications for repellency, flame retardance, soil release, and static control was covered in an extensive review [8].

Whatever the reasons for modification, the strategies can be grouped under the following headings.

7.5.1 Include co-monomers in the polymerization process

The extent to which this can be accomplished without fundamentally changing the nature of the polymer in its definition or other properties is debatable. Some basic criteria must be met in order for the material to be satisfactory: the co-monomer must be stable during polymerization and extrusion and not lead to a polymer with markedly inferior mechanical properties.

Thus, for example, challenges in dyeing polyester have been answered in part by the development and commercialization of PET that includes a proportion of 5-sulfoisophthalic acid as a co-monomer: the resulting sulfonic acid groups provide "sites" for dyeing with cationic ("basic") dyes [9]. Co-polyesters of PET have been examined for obtaining deep dyeing material [10]. For flame retardance, bromine-containing co-monomers have been developed, such as 2,5 dibromophthalic acid [11]. More recently, phosphorus-based co-monomers have been examined [12].

In general, the use of a co-monomer as opposed to an additive in the melt results in reduced leaching tendency in later use. This may be an advantage or disadvantage. Related to the use of a co-monomer is the inclusion of some agent that affects the DP or linearity of the polymer. Thus, the inclusion of a depolymerization agent in the melt produces a PET with a lower MW (narrower weight distribution) that pills less [13], and the addition of 6–16% (by weight) of polyethylene glycol together with ca. 0.1% pentaerythritol branching agent increases the wetting and wicking behavior of the polyester fiber [14].

7.5.2 Include additives to the polymer melt before extrusion

Additives to the melt become embedded in the fiber structure after extrusion and drawing. The additive must be stable to the temperature of the melt, i.e., up to about 300°C. The additive must be readily miscible with the melt and not significantly change its rheological properties. It must also not degrade the polymer at the high temperatures. The resulting polyester should not have markedly worse physical properties. Subsequent use should also not remove the additive, unless leaching is a required property.

The prime example in this case is the routine inclusion of titanium dioxide as a delustrant. Colorants can be added as the fiber is formed, and around 5% of the total production of polyester is thus produced as "solution dyed" material. Flame retardancy has also been achieved using additives to the melt, especially bromine [15, 16] and phosphorus compounds [17]. To avoid problems of interaction when antimony compounds are used as adjuncts, blending of melts, skin-core spinning, and microencapsulation of the additive have been suggested [18, 19]. Static may be controlled by incorporating carbon into the fiber matrix [20], and the improved compatibility of polyester with a carbon bicomponent has been described [21]. When a surface effect is required, this method, like co-polymerization, is of limited value, since the bulk nature of the fiber is being altered. It is thus not usually used to affect the hydrophilicity (hence soil release or moisture transport properties) of the fiber.

7.5.3 Topical additive treatments after the fiber is formed

Chemical additives to the solid fiber may be divided into polymeric materials that lie on the surface, and do not penetrate to any significant extent, and smaller moieties that rely on some measure of fibrophilic penetration into the polyester matrix.

7.5.3.1 Surface polymer additives

Polymeric additives that lie on the surface of the polymer may be applied as a melt, solution, or emulsion. The durability is enhanced if the polymer is cross-linked as part of a curing or fixing process. Further, durability enhancement occurs if some bonding takes place between the polymer and the fiber. Given the paucity of functionality on the surface of polyester, this may involve a pre-treatment to create functionality or occur *via* transesterification reactions.

Soil release finishes can be grafted, transesterified, or generated by sorption of surfactants at the surface [22]. Repellency from silicones has been widely studied, and increasing durability is obtained by cross-linking them or by incorporating groups that react with the substrate. They can also be grafted and linked by this reaction [23, 24]. Perfluoroalkyl acrylic polymers are increasingly common [25, 26] and provide both oil and water repellency. So-called "dual acting" fluorochemicals that additionally contain oxyethylene segments can provide both soil release and soil repellency [27]. Soil release can be obtained by the application of vinyl polymers containing acrylic acid [28]: if a cross-linker is included, durability increases [29]. The anionic hydrophilicity may be augmented by non-ionic oxyethylene moieties.

Surface polymer deposits at a PET surface can be generated by "grafting" in which a monomer is applied and polymerization induced by the generation of initiation. Grafting can be performed during the activation process (direct irradiation method) or on a pre-activated surface (post-irradiation method) [30]. UV-induced graft polymerization of water-soluble monomers has been used to increase the hydrophilicity of PET permanently without changing its bulk properties [31]. The anti-static properties and wicking time were also improved. In a similar study, a thin functional layer was established at a PET surface by cross-linking reactive substances by irradiation with excimer UV-lamps. Dyeability was also improved [32]. Laser treatments are used in grafting: a CO_2 pulsed excimer laser in air formed peroxides that initiated graft co-polymerization of acrylamide on the surface of PET film to improve wettability [33]. Vinyl compounds containing bromine or phosphorus have been grafted onto polyester to provide flame retardance [34]. The grafting can be made deeper by pre-swelling with ethylene dichloride. Repellency requires surface treatment, and in routine use, zirconium and aluminum soaps, combined with wax emulsions, have been used for many years [35]. Durable anti-static finishes based on polyamines modified again with oxyethylene segments have been widely used [36].

7.5.3.2 Fibrophilic, "dyeing" additives

The modifying substance, usually non-polymeric, or at least a low polymer, may be diffused into the polymer matrix. This is akin to dyeing and relies on a measure of substantivity. The dyeing may be achieved by exhaustion from an aqueous bath or by a pad heat ("thermosol") technique. Effects may be confined to the surface by adjustment of the application conditions, especially if the molecular make-up of the diffusing substance has "fibrophilic" and fibrophobic portions: the fibrophilic part sinks into the matrix while the active, non-fibrophilic part remains at the surface. The diffusion into the polymer matrix can also be limited by the time and temperature at which the diffusion takes place.

Thus for flame retardancy, the now discredited "tris" (*tris*-2,3-dibromopropyl phosphate) was exhausted or dyed to give 4% add-ons [37]. Soil repellency is similarly a surface phenomenon and is important in the case of polyester. In an elegant piece of work [38], increasing the temperature of application increased the sorption of the finish to anchor a lipophile and leave a hydrophilic segment protruding from the surface. A polyester emulsion co-polymer with some ethylene, and some oxyethylene segments can be pad heated to provide soil release [39]. Other oxyethylene moieties can be applied together with dyestuffs in an exhaust procedure [40]. Non-ionic surfactants cloud up at high temperature and deposit on the fiber surface, or are absorbed in a dye-like manner.

7.5.4 Physical/non-additive chemical modifications

Several other approaches do not involve the use of finishes *per se*, but instead rely on chemical reactions to modify the polyester itself. The foremost of these, alkaline hydrolysis, has been widely studied. Aminolysis is less widely known. The use of solvents as swelling agents or that other wise promote diffusion into the fiber has been widely studied for improving the dye uptake of polyester, but other than that has been comparatively little studied. Steam explosion is in its infancy, and laser treatments have been examined in several studies. More esoteric but better established are modifications based on electrons, ions, and other transient reactive species that are present in plasmas, flames, and coronae.

7.5.4.1 Alkaline hydrolysis

Polyester may be made more hydrophilic by alkaline hydrolysis. This has been widely studied and the work extensively reviewed [41–43]. Hydrolysis of PET can occur under acidic conditions or under alkaline conditions. Acid hydrolysis is not a useful or practical process for modification of polyester. Under alkaline conditions, the carbonyl oxygen atom of the ester group is attacked by the hydroxyl anion to produce one hydroxyl and one carboxylate end group (see Figure 7.2). The rate of diffusion of the alkaline reagent determines the rate of hydrolysis. The hydrolysis of polyester groups initially causes (surface) chain shortening, and the generation of hydrophilic/functional carboxylic acid and hydroxy groups. As the chains become shorter, continued reaction will result in the loss of material and a reduction of fiber diameter, but the overall molecular weight remains largely unchanged. This is used commercially as the so-called "denier reduction". If carried out extensively, hydrolysis results in

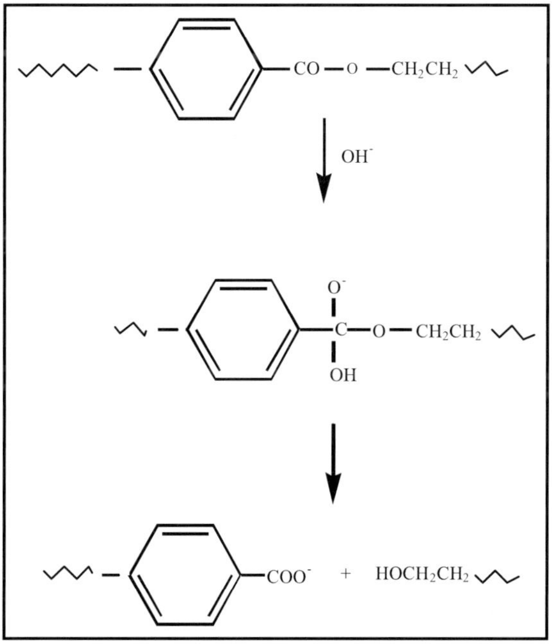

Figure 7.2. Alkaline hydrolysis of PET

loss of strength and weight, and care is needed to make sure that maximum strength is retained. The use of enzymes for this hydrolysis has recently been suggested [44].

The use of a quaternary ammonium salt to augment the alkali has been widely suggested: one recent example has examined its use for modifying the strength of non-woven materials [45].

7.5.4.2 Aminolysis

Several studies have assessed the effects of amine interaction with polyester. Early studies assessed the aminolysis of polyester as a means of examining fiber structure without regard to maintaining the integrity of the polymer [46, 47]. The degradation effects on polyester of a monofunctional amine versus alkaline hydrolysis have been studied [48]. These studies, which again involved high levels of fiber degradation, demonstrated that alkaline hydrolysis has a more substantial effect on fiber weight without extensive strength loss. In contrast, aminolysis had less effect on fiber weight but decreased fiber strength, indicative of a reaction within the polymer structure rather than simply at the surface. It was later demonstrated that bifunctional amine compounds could be reacted with the polymer with minimal loss in strength while generating amine groups at the fiber surface [49]. The early stages of the reaction were largely confined to the fiber surface and the resulting fiber had modified wetting properties and improved adhesion with the matrix when used in composites.

A recent paper by our research group has re-examined the interaction of untreated and alkali hydrolyzed polyester with a range of aliphatic diamines [50]. 1,6-Hexanediamine, 2-methylpentamethylene diamine, 1,2-diaminocyclohexane, tetraethylenepentamine, and ethylene diamine were applied to untreated polyester. Ethylene diamine was also applied from a range of solution concentrations in toluene. The treatment generated amine groups on the fiber surface and was revealed by staining with anionic dyes under conditions in which the amine group was protonated. Unexpectedly, the reaction resulted in the simultaneous formation of carboxylic acid groups in a manner similar to alkaline hydrolysis, revealed by staining with Methylene Blue (Figure 7.3). The reaction thus resulted in a bifunctional polyester surface. The ratio of amine and carboxylic acid groups differed with unhydrolyzed and hydrolyzed starting materials (Figure 7.4). Strength loss was somewhat greater than with alkaline hydrolysis.

7.5.4.3 Solvent swelling

Before the widespread introduction of pressurized dyeing machines, emulsified solvent-type materials were widely examined and the better ones widely used as "carriers" to allow satisfactory polyester dyeing [51]. These materials

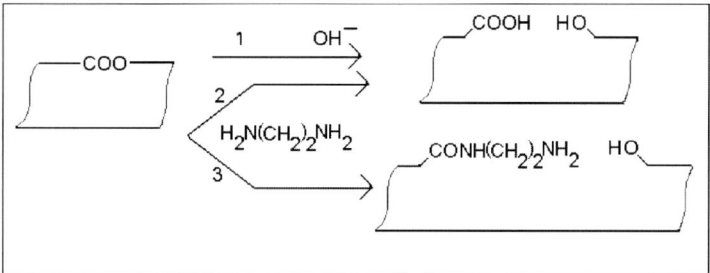

Figure 7.3. Hydrolysis and aminolysis of polyester

effectively lowered the glass transition temperature of the fiber. The range of more general solvent-type pre-treatments that produce improved dyeability has been reviewed, and the effect on a range of fiber properties is included [52].

7.5.4.4 Steam explosion

The process of steam explosion involves exposing a sample to conditions of high pressure and temperature for a certain period or residence time, and then explosively discharging the product to atmospheric pressure [53]. The rapid expansion of any water molecules present in the material occurs over a very short time and hence causes changes at the morphological level only [54]. These two effects increase the specific surface area, accessibility, and hence, the reactivity of the material [55]. The application of steam explosion technology to PET has recently been examined [56]. Steam-exploded PET fabrics were

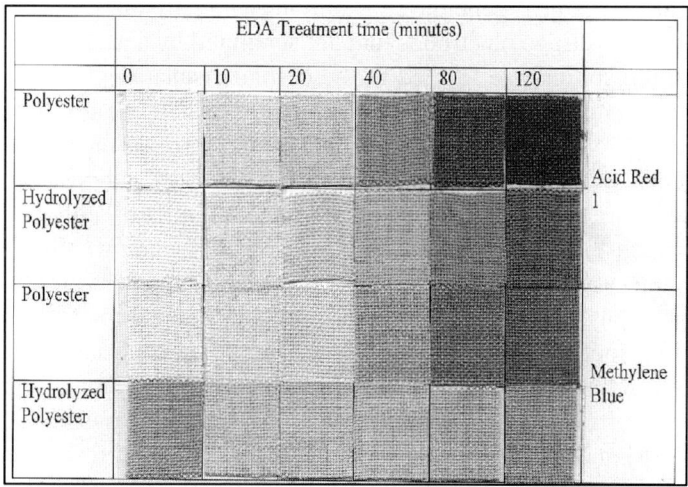

Figure 7.4. Staining of EDA-treated, hydrolyzed, and unhydrolyzed polyester

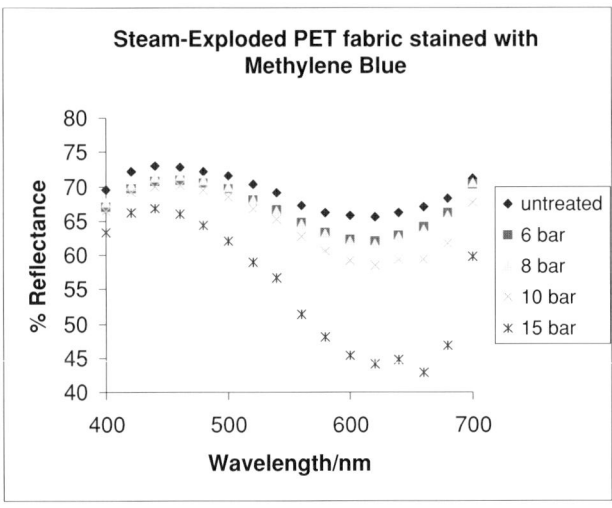

Figure 7.5. Reflectance values of steam exploded PET fabric, stained with methylene blue

stained with Methylene Blue (C.I. Basic Blue 9) dye to determine the presence of acidic functional groups on the surface of the fabric as a result of treatment. The reflectance values of the stained fabrics and of the unstained equivalent are illustrated in Figure 7.5. The decrease in reflectance at ca. 600 nm corresponds to increased staining. There is a gradual increase in the effect of treatment with increasing pressure up to 10 bar: at 15 bar, there is a much larger increase in the effect of treatment.

7.5.4.5 Corona treatments

A corona discharge is generated by applying a high frequency voltage, under atmospheric pressure to electrodes that are separated by a narrow gap. Ozone, radicals, and electrons are formed in the corona discharge [57]. Background radiation generates a small number of electrons that cause the formation of these highly active species [58]. "Corona" is the term given to this mixture of gas molecules, excited molecules, electrons, radicals, and positive ions that are formed between the electrodes, and coronae will modify polymer surfaces passed through the electrode gap. This is a useful technique since it can be operated in air at atmospheric pressure [59]. The changes that occur are both chemical and physical. Polar functional groups are introduced to the surface by oxidation, increasing the wettability of the polymer. Changes in surface morphology of polymeric films, caused by electron bombardment, improve the adhesion properties of certain polymers. Corona treatment has been evaluated for olefin [58] and wool [59], but the method has received little attention for use on PET.

7.5.4.6 Excimer laser treatments

Pulsed excimer lasers first became commercially available in the late 1970s [60]. The name excimer was derived from the term "excited dimer", referring to a group of diatomic molecules that are chemically stable in the excited state rather than the ground state. The surface of synthetic fibers can be modified by irradiation with a pulsed excimer laser. Laser irradiation penetrates only a few micrometers of the surface without interfering with the interior structure. Photodissociation and ablation of the polymer occur as a result of treatment with a UV pulsed excimer laser [61].

Such UV irradiation has been carried out on PET fabric [62]. The UV irradiation penetrated the top layer of the surface only. SEM micrographs showed the rippled structures on the surface of the fabrics due to laser treatments. A decrease in surface oxygen content and the formation of a thin carbonaceous layer on the PET surface increased the contact angle of PET treated with a high-energy dose, i.e., the treated PET was more hydrophobic. Lower energy pulses caused a decrease in contact angle. The effect was stable for at least 1 month.

Compared to these conventional excimer lasers, frequency-doubled copper vapor laser (FDCVL) systems are cheaper and safer for modifying polyester fibers and films. The pulse energies obtained from FDCVLs are less than those obtained from excimer lasers, but repetition frequencies of up to 20 kHz result in higher average power levels and minimal peripheral damage. FDCVLs also possess a relatively high spatial coherence and low divergence. An experiment used an FDCVL to scan a beam across the surface of warp knit PET microfiber fabrics and films repeatedly at different speeds and scan times. Even short exposure times improved the surface functionality of PET fabrics and films [63].

7.5.4.7 Flame treatments

Flame treatments were developed initially in the 1950s to increase the wettability of polyolefin films. More recently, flame treatments have been used in the modification of paperboard and of thicker polyolefin materials, for example, automobile body parts [64]. Corona treatments have generally played the major role in modification of films due to the ease of operation. Flame treatments are currently being revived in the modification of films, as concerns about safety and efficiency of the technique have been improved. Another advantage of flame treatments is that the improved properties of the flame-treated surfaces, such as increased wettability do not wear off over reasonable times, whereas corona-treated surfaces deactivate quite rapidly [65]. As with corona treatments, this technique remains open to exploration on polyester.

7.5.4.8 Plasma treatments

Plasma is a result of ionization, fragmentation, and excitation processes. Electrons colliding with gaseous molecules form excited molecules, ions, and energetic photons [66, 67]. These excited species react further to form charged particles, radicals, and UV photons. This mixture of charged particles is extremely reactive with surfaces that are exposed to the plasma [68]. Ions generated in the plasma are accelerated to energies of 50–1000 eV, causing sputtering or etching of thin films [69]. Given a continuous supply of energy to maintain the plasma state, the highly energetic, active species within the plasma will create new macromolecular surface structures [70]. They do so while having minimal or no effect on the bulk properties [70]. The increase in surface energy brought about by plasma treatments results in an increase in wettability and can improve dyeing, moisture uptake, and fabric handle.

Non-polymerizing plasma, such as oxygen or nitrogen, etches and chemically modifies the surface of the polymer substrate. The extent of the modification depends on the substrate, the type of gas used, the treatment time, gas pressure, and RF power. Fibers or fabrics experience a mass loss due to the etching of the topmost layer of the surface by bond scission, and etching or sputtering [71]. Such etching occurs preferentially in the non-crystalline regions of the polymer surface. An examination of the ultimate dyeability of PET and nylon 66 fibers of varying crystallinity exposed to plasma etching confirmed that the attack was concentrated on the non-crystalline regions [72]. Oxygen plasma can increase the wettability of the fabric by introducing polar functional groups onto the surface. Nitrogen plasmas introduce $-NH_2$ groups that provide extra dyesites onto the fabric, increasing its dyeability. A hydrogen/nitrogen plasma mixture generates free carbon radicals on the surface that form carbon–carbon cross-links on the surface and reduced dyeability. An inert gas such as argon can cause sputtering on the material being treated. Chemical etching during plasma can occur when using an oxidative gas, such as oxygen [73]. Nitrogen and oxygen plasmas have been shown to provide PET with reduced wetting times and the most durable wettability by the production of fine micropores on the surface, whereas ammonia plasmas did not [74].

The introduction of functional groups onto a plasma-treated surface is known as implantation. Activated gas molecules from the plasma interact with the polymer surface to implant new functional groups at the polymer surface. The type of functional group introduced depends on the type of gas used [71], whereas the number of free radical species generated in the plasma will depend on the substrate. Various low-temperature plasmas, including O_2, N_2, H_2, and Ar, were applied to cotton, linen, PET, nylon, wool, and silk [75]. The free-radical yield was greatest for the cotton fabric, followed by wool, silk, nylon, and PET. More stable free radicals were formed in the natural fibers. Various studies have been carried out to introduce hydrophilic functional groups onto the surface of PET, *via* plasma irradiation, to improve its wettability and dye-ability [76]. RF-plasma-generated sillylium cations can be used to activate inert polymeric surfaces [77, 78]. The $SiCl_3^+$ ions generated in the plasma are readily converted into $Si(OH)_x$ functionalities when in contact with moisture, allowing PET to be dyed with basic dyestuffs.

The immediate effects of plasma treatments wear away over time. The adsorption of contaminants from the atmosphere occurs very rapidly. Surface reorganization occurs slowly over time, and low molecular weight material formed from the plasma treatments can diffuse into the bulk material: as it does so the surface properties of the plasma-treated material will change [79].

Comparison of various gas-phase processes for the surface modification of polymers

Corona, flame, and plasma treatments are all used to generate very fast surface oxidation. UV and ozone treatments oxidize the surface at a much slower pace. The treatments on the surface of PP and PET films have been compared [64].

Corona, plasma, and flame treatments oxidized the polymer surface to the target level in <0.5 s, but much longer times were required to oxidize the polymer surface using UV/ozone treatments. The researchers also discovered that the contact angle measurements for each technique varied suggesting that the surface chemistry derived from each technique is slightly different. Flame-treated samples were the most wettable, as most of the oxidation occurs in the outermost 2–3 nm. FTIR analysis shows that UV/ozone treatments can modify the polymer surface to a greater depth.

Long exposure corona treatments of PE films reduce the peel strength of the film being treated due to an increased formation of low molecular weight oxidized material: this is not the case for plasma treatments [80]. Obviously, long exposure treatments are not an option for flame treatments. Corona treatments have been more widespread in industry than plasma treatments because industrial scale plasma equipment is more expensive and requires a vacuum. However, corona treatments are limited to one-dimensional material, whereas plasma treatments can modify three-dimensional material readily. Flame treatments can also modify three-dimensional materials but are limited to treating materials that cannot be thermally damaged. Plasma, however, is capable of modifying heat-sensitive polymeric materials [64].

Plasma-induced grafting

Grafting was referred to earlier. Plasmas can be used to induce polymerization of monomers at the fiber surface. Monomers can be pre-selected to produce the desired surface property onto the treated polymer, and the treatment is stable over time. Peroxide-initiated grafting occurs on immersing an oxygen- or air-plasma-treated polymer into a solution of monomer, with a high yield of grafted monomer on the surface [81]. Among the many reports on this technique is included the modification of polyester [82]. Graft fluorination of PET fiber was used to produce oil-resistant and water-repellent materials [83]. The authors found that a hydrophobic surface was reached at a grafting yield of 3.65%. Exposure of a PET film to argon plasma, followed by the introduction of acrylic acid vapor into the plasma chamber, generated a more wettable surface than a liquid-phase grafting [84].

7.6 Limitations of polyester in medical use

Polyester in medical use has undoubtedly improved the quality of life for an aging patient population. However, all implantable devices are prone to three major complications: surface thrombus formation, post-surgical/wound bleeding, and incomplete/non-specific cellular healing.

Surface thrombus formation is the result of the interaction of blood with the relatively thrombogenic biomaterial surface. This problem is evident in medium/small-diameter vascular grafts as well as stents and catheters, which can fail early after implantation due to occlusive thrombus formation within the blood-contacting surface of the device.

A complication of all implantable biomaterials is incompatibility between flowing blood and the biomaterial surface. Thrombin, a pivotal enzyme in the blood coagulation cascade, has been implicated as the primary agent responsible for thrombus formation. The initial interaction of blood and the foreign surface results in a myriad of responses: platelet activation and adhesion [85], activation of the intrinsic pathway of the coagulation cascade resulting in formation of active thrombin [86], leukocyte activation [85], and the release of complement and kallikrein [87]. If unregulated, these responses lead to mural thrombus formation with subsequent occlusive thrombosis and failure of the implanted biomaterial.

Uncontrolled post-surgical/wound bleeding is directly related to the limited interaction of blood with the foreign surface as well as the physical construct (i.e., porosity, weave design) of the material. For hernia repair mesh and wound dressings, the problem arises in that the overall time to create surface thrombus is extensive, thereby delaying hemostasis and compromising the patient.

Lastly, incomplete/non-specific cellular healing affects various medical devices. Since biomaterials are composed of foreign polymeric materials, cellular components normally present within native tissue are not available for the reparative process. These complications are evident in medical devices such as vascular grafts, hernia repair mesh, wound dressings, and catheter cuffs [88–90].

Common to these complications is that currently available biomaterials do not emulate the multitude of dynamic biologic and reparative processes that occur in normal tissue. Thus, development of a novel biomaterial that would emulate some of the normal healing processes of native tissue would improve patient morbidity and mortality upon implantation of various medical devices. Exhaustive studies have been aimed at creating a novel biomaterial surface by either non-specific binding of a biologically active agent, covalent linkage of an agent with a broad spectrum of activity, or altering the biomaterial surface. Thus far, none of these technologies have resulted in a clinically used biomaterial surface.

7.7 Efforts to combat limitations in medical use

A survey of the broad range of attempts to overcome these shortcomings for vascular grafts has been published, and the reader is referred to that publication [91]. What follows extends and complements that work.

Medical limitations have some direct correlations with limitations in normal use. Where a surface interacts with a biological system, many factors relating to the nature of the surface can modify the biological response. In many cases, the interaction is unpredictable, and there is much still to learn about the effect of the physical and chemical nature of the surface on these interactions. The techniques for modification in normal use discussed earlier thus form a rich basis for exploratory work in providing better materials for medical use. More specifically, the modifications that generate surface functional groups allow covalent binding of specific proteins. Where more than one functional group is present, the possibilities exist for the binding of more than one bioactive moiety. Our research group has been much involved in producing a more successful medical polyester material. Using the techniques outlined in section 5.3.2, we have "dyed" polyester with antibiotics [92]. We have also looked to link bioactive proteins to the surface of polyester and have found that carboxylic acid groups generated by alkaline hydrolysis (section 5.4.1) are useful in such covalent linking [93]. We have linked clot-preventing proteins and a growth factor to promote cellular ingrowth and demonstrated that they retain their activity when linked [94, 95].

7.7.1 Combating surface thrombus formation

Many attempts to create a more biocompatible surface have been based on establishing a new biologic lining on the luminal surface that would "passivate" this initial clotting reaction. These have ranged from non-specifically binding albumin to the surface followed by heat denaturation [96] to non-specifically cross-linking albumin [97], gelatin [98], and collagen [99]. Covalent or ionic binding of the anti-coagulant heparin alone [100], in conjunction with other biologic compounds [101] or with spacer moieties [102], and covalent link-age of thrombomodulin [103] have also been performed. Other studies have focused on modifying the composition of the biomaterial by either increasing hydrophilicity *via* incorporation of polyethylene oxide groups [104] or creating an ionically charged surface [105].

The success of these approaches has been limited: (1) thrombin is not directly inhibited, therefore fibrinogen amounts remain constant on the mate-rial surface permitting platelet adhesion; (2) heparin-coated biomaterials may be subject to heparitinases limiting long-term use of these materials; (3) non-specifically bound compounds are stripped from the surface under the shear stresses of blood flow, thereby re-exposing the thrombogenic biomaterial sur-face; (4) rapid release of non-specifically bound compounds may create an undesired systemic effect; and (5) charge-based polymers may be covered by other blood proteins such that anti-coagulant effects are masked.

7.7.2 Post-surgical/wound bleeding

Incompatibility between blood and the biomaterial surface complicates the use of all implantable biomaterials. The initial interaction of blood and surface results in a myriad of responses: platelet activation and adhesion [106], activation of the intrinsic pathway of the coagulation cascade resulting in formation of active thrombin [107], leukocyte activation, and the release of complement and kallikrein [108]. While this response is desirable for hernia repair mesh, wound dressings, and catheter cuffs, the intensity and overall rate at which this occurs are limited. Additionally, the physical construct (i.e., porosity, weave design) of the material affects the rate and the extent at which thrombus formation occurs. For hernia repair mesh and wound dressings, if the overall time to create surface thrombus is extensive, hemostasis is delayed and the patient may be compromised.

7.7.3 Promotion of cellular adhesion/growth

The development of a uniform cellular layer on the implanted biomate-rial has been proposed to enhance biocompatibility. These cells, while provid-ing structural stability *via* material incorporation into the surrounding tissue, maintain hemostasis, prevent infection, and synthesize bioactive mediators. This type of cellular incorporation does not occur in actuality, thereby pre-disposing these biomaterials to infection [109, 110] and thrombosis [111, 112]. Thus, failure of appropriate cell-type growth and development to these bioma-terials significantly limits their expanded use.

Cellular adhesion to biomaterials using cell-seeding techniques has been extensively employed [113, 114]. Adhesive proteins such as fibronectin, fibrinogen, vitronectin, and collagen have served well in graft-seeding protocols [115]. The cell-attachment properties of these matrices can also be duplicated by short peptide sequences such as RGD (Arg-Gly-Asp) [116]. These adhesive proteins, however, have several drawbacks: (1) bacterial pathogens recognize and bind to these sequences [117]; (2) non-endothelial cell lines also bind to these sequences [118]; (3) patients requiring a seeded material such as a vascular graft have few donor endothelial cells, therefore cells must be grown in culture [119]; and (4) endothelial cell loss to shear forces from flowing blood remains a significant obstacle [120].

Modification of the surface has also been employed to modify host response to the foreign body, serving as an approach for improving cellular adherence. Cells that have been seeded have been shown to attach and grow better on a variety of protein substrates coated onto the biomaterial [121]. Bioactive oligopeptides [122] and cell-growth factors [123] have been immobilized onto various polymers and shown to effect cell adherence and growth. Additional studies have described the incorporation of growth factors into a degradable protein mesh, resulting in the formation of capillaries into the material [124]. Utilizing these techniques to incorporate growth factors, however, does have limitations: (1) growth factor is rapidly released from the matrix; (2) matrix degradation re-exposes the thrombogenic surface, thus endothelialization is not uniform; and (3) release of non-endothelial specific growth factor is not confined to the biomaterial matrix, thereby exposing the "normal" distal artery to the growth factor.

7.7.4 Use of functional groups for protein attachment

Many of the modification methods discussed (hydrolysis, aminolysis, plasma, laser, etc.) generate surface functional groups in much larger amounts than are present on original polyester. These functional groups can be the basis for covalent attachment of a variety of bioactive agents. Earlier work to bind proteins to the functional groups created by alkaline hydrolysis has been referred to earlier. This work demonstrated that bound bioactive proteins maintain their activity once bound. More recent work has examined the use of functional groups derived from those other modification techniques and also has begun work to generate a multifunctional surface *via* the binding of different proteins or a combination of protein binding with infection resistance derived from the "dyeing" of antibiotics.

Covalent linkage of a protein to a biomaterial surface in order to create a "basecoat" layer has numerous advantages. Such a layer has been shown to "passivate" a surface that is relatively thrombogenic, thereby decreasing adhesion of blood products such as platelets, red blood cells, and fibrinogen [96]. Proteins incorporated as a basecoat layer have been used as "scaffolding" in order to promote a specific response such as linkage of RGD peptides to promote cell adhesion [125]. Additionally, Park et al. [126] demonstrated that increasing the angstrom distance between a biologically active molecule and the surface *via* polyethylene oxide groups reduced steric hindrance on the target molecule, thereby maintaining activity. Covalent linkage of a protein "basecoat" layer would serve as the spacer between the biologically active moiety and the surface. Albumin, which is in natural abundance in circulating blood, has shown numerous beneficial results *in vitro* and *in vivo* in earlier studies [127, 132] and was used in this study. Utilization of a basecoat layer could also permit significant amplification of potential binding sites for secondary protein attachment *via* heterobifunctional cross-linkers, creating a biomaterial surface with distinct properties for a specific application. This BSA surface has been shown to possess numerous binding sites for other biologically active proteins such as recombinant hirudin (rHir) [128].

7.7.4.1 *Protein binding to hydrolyzed polyester*

In work in our laboratories, BSA was radiolabeled using ^{125}I. Following iodination, bound ^{125}I was determined. These values were then utilized to derive a specific activity that was used to determine protein bound to the material surface. Samples of polyester were scoured with and without hydrolysis. BSA was then bound to the hydrolyzed polyester surface using heterobifunctional carbodiimide cross-linker EDC. Control and another set of hydrolyzed segments were reacted under the same parameters without EDC cross-linker. Control and test segments were then reacted with the ^{125}I-BSA solution for 2 h at room temperature. The specimens were sonicated three times in detergent solution to remove any weakly adherent ^{125}I-BSA, then gamma counted to determine the amount of ^{125}I-BSA covalently bound. The amount of ^{125}I-BSA covalently bound to the hydrolyzed segments *via* EDC was significantly greater than both control and hydrolyzed controls at all solvent concentrations evaluated (Figure 7.6).

7.7.4.2 *Protein binding to a bifunctional polyester surface*

A similar technique was used to immobilize protein to a bifunctionalized polyester surface. Scoured control and hydrolyzed segments were reacted with ethylenediamine. The EDA-exposed segments were then removed and placed into distilled water overnight at room temperature (C-EDA or Hyd-EDA) [134]. Staining with Methylene Blue and with C.I. Acid Red 1 was used to quantify the two functional groups present. The C-EDA segments were then assessed for ^{125}I-BSA binding *via* a range of commercially available heterobifunctional cross-linkers (Traut's reagent, sulfo-SMCC, and sulfo-SPDP). These

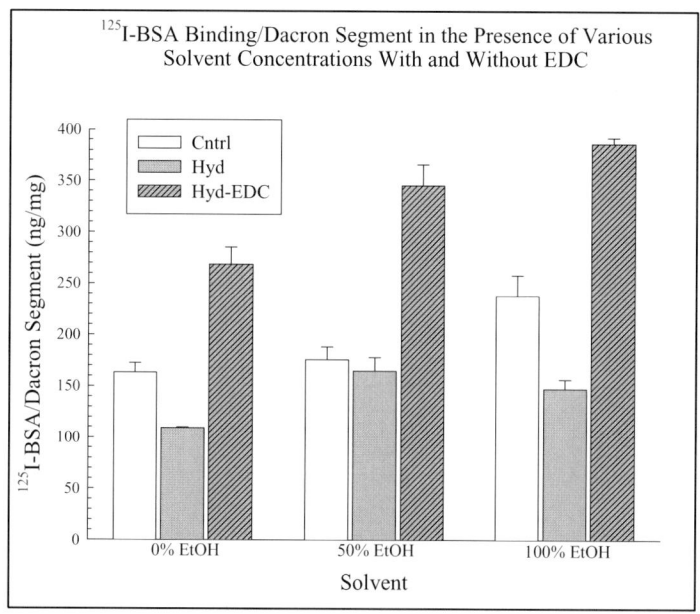

Figure 7.6. Protein binding to hydrolyzed polyester

cross-linkers react specifically with primary amine groups, but vary in solubility and chain length. EDA-treated PET was reacted with the various cross-linkers and washed. Sulfo-SMCC was added to ^{125}I-BSA solution and reacted for 20 min at 37°C in a water bath. The ^{125}I-BSA–SMCC intermediate was then purified *via* gel filtration. The ^{125}I-BSA–SMCC solution was then added and reacted for either 3 or 20 h at room temperature on an orbital shaker (150 r.p.m.). After incubation, control and test segments were removed and washed four times in PBS with Tween 20. Segments were then gamma counted. Using protein concentration determined *via* Lowry assay and gamma counts of a set ^{125}I-BSA volume (i.e., specific activity), the amount of ^{125}I-BSA/Polyester segments was determined. C-EDA segments incubated with the various cross-linkers had significantly greater ^{125}I-BSA binding when compared with C-EDA exposed to protein only (non-specifically bound segments) materials (Figure 7.7). The linking combination of Traut's reagent with sulfo-SMCC resulted in the binding of 604 ng ^{125}I-BSA per 1 mg polyester, regardless of the cross-linker reaction (surface reaction versus protein reaction). Reaction of SPDP (surface) with sulfo-SMCC (protein) resulted in lower protein binding than the Traut's–sulfo-SMCC reaction. Increasing the reaction time from 3 to 20 h not only increased total protein binding, but also significantly increased non-specific binding. Reaction of the surface with sulfo-SPDP (surface) with Traut's (protein) for 20 h had the greatest ^{125}I-BSA binding. Non-specific binding also increased.

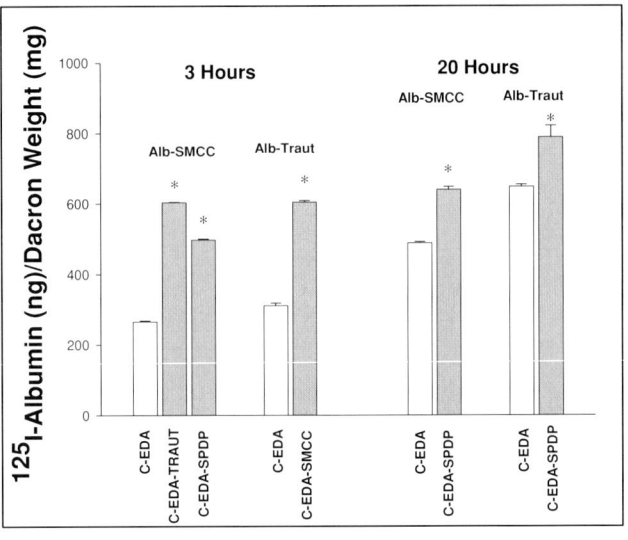

Figure 7.7. Protein binding to a bifunctional polyester surface

7.7.4.3 Protein binding to laser-treated polyester

The linking of protein to carboxylic acid groups created *via* laser exposure was assessed. Laser hydrolysis permits localized formation of carboxylic acid groups, whereas sodium hydroxide hydrolysis forms carboxylic acid groups through the entire material. The laser treatment was varied: the control software was set up to scan knitted polyester segments (30 mm × 30 mm) at a scanning speeds of x (mm/s) for a total scan cycle time of y (s). After scanning, the samples were washed in a mild detergent solution, rinsed, and air-dried. As before, methylene blue staining was used to quantify carboxylic acid group formation. Knitted and woven polyesters, scoured or scoured and sodium hydroxide-hydrolyzed, were used as controls. Polyester specimens were immersed into EDC solution in 100% ethanol for 30 min at room temperature, then removed, shaken, and placed into a ^{125}I-BSA solution for 2 h at room temperature. The segments were removed then sonicated in PBS containing Tween 20 for 5 min. The amount of ^{125}I-BSA bound (ng) per 1 mg polyester was calculated for each segment.

Carboxylic acid formation was evident from the uptake of Methylene Blue, with formation increasing with increasing laser exposure time. Modification was limited to the specific area exposed to the laser. EDC reaction with all materials containing carboxylic acid groups resulted in significantly greater ^{125}I-BSA binding (Figure 7.8). For the laser-treated knitted polyester, ^{125}I-BSA binding was slightly less than the knitted hydrolyzed polyester (740 ng ^{125}I-BSA per 1 mg polyester) but is significant given the limited area of the material modified. Non-specific ^{125}I-BSA binding to the laser-treated and hydrolyzed

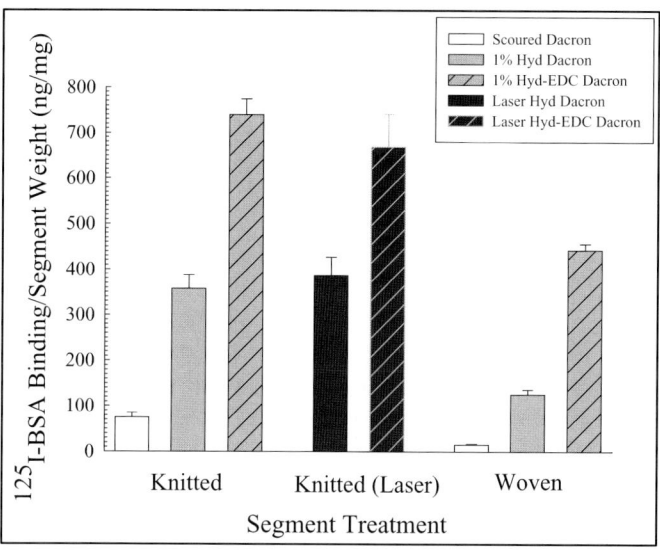

Figure 7.8. Protein binding to a laser-treated polyester surface

polyester surfaces was similar. ^{125}I-BSA binding to the woven segments was comparable to that observed in the earlier studies.

7.7.4.4 Development of an anti-microbial bioactive polyester surface

The earlier-mentioned studies evaluating protein immobilization to the various functional groups served as the foundation for creating a multifunctional surface. In our previous studies using unmodified polyester, broad-spectrum fluoroquinolone antibiotic Ciprofloxacin (Cipro) uptake into the fibers was unsuccessful using exhaust dyeing, with anti-microbial activity lasting <4 h after extensive washing [129–131]. Pad heating, a high temperature technique that opens the fiber structure, was used, and treated polyester demonstrated anti-microbial activity for >50 days. While this technique was successful for incorporating Cipro, the elevated temperatures are not conducive to maintaining the bifunctional groups created on the material surface (data not shown). In order to develop infection resistance within a modified polyester material, exhaust dyeing of the Cipro onto this more hydrophilic surface was re-examined. EDA-treated polyester segments were placed into a Cipro "dyebath" of liquor ratio 20:1, 5% owf Cipro pH $= 8.0$, dyeing for 2 h at 70°C. After air-drying overnight, Cipro-dyed segments were autoclaved for 15 min (10 min dry). Control segments were treated in a similar fashion; however, no heating or autoclaving was performed. Segments were also stained to determine the presence of the functional groups post-dyeing. Macroscopically, a yellowish hue was evident after Cipro-dyeing into the C-EDA segments. C-EDA segments dipped into

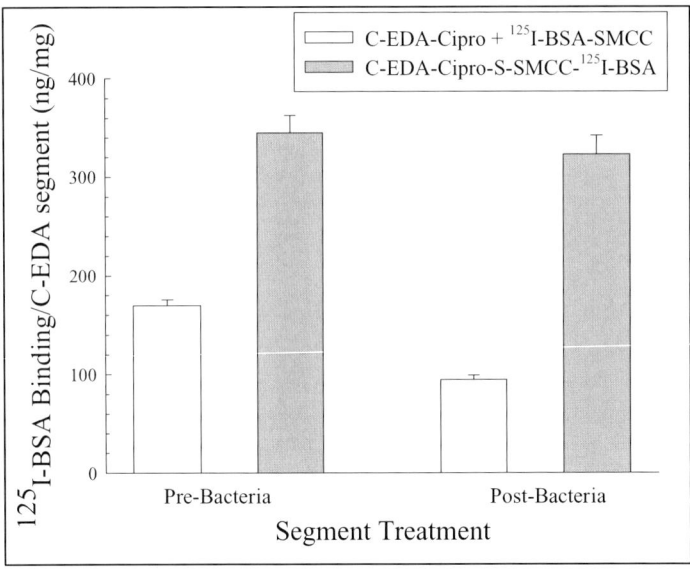

Figure 7.9. Protein binding to an antibiotic-modified polyester pre- and post-exposure to bacteria

Cipro but not heated had no gross color change. Exposure of pre- and post-dyed C-EDA segments to staining had no visible difference in dye uptake demonstrating that the functional groups were not affected by the low-temperature dyeing.

Cipro-dyed C-EDA and control segments were treated with Traut's reagent for 1 h and then washed twice with bicarbonate buffer. Sulfo-SMCC was added to a ^{125}I-BSA solution and reacted for 20 min at 37°C. The ^{125}I-BSA–SMCC intermediate was purified *via* gel filtration then added reacted for 3 h at room temperature with the C-ED-Traut's segments. Segments were removed and washed in PBS with Tween 20 for 5 min. Using protein concentration determined *via* Lowry assay and gamma counts of a set ^{125}I-BSA volume (i.e., specific activity), the amount of ^{125}I-BSA (ng)/bD-PU-AB segment (mg) was determined.

Incubation of the Cipro-dyed C-EDA segments with Traut's reagent resulted in two-fold greater ^{125}I-BSA binding compared with controls without Traut's (Figure 7.9). Even after exposure to bacteria, ^{125}I-BSA remained bound to the C-EDA surface. Sonication is typically used to remove non-specific protein binding. However, a combination of the wash buffer and sonication was shown to remove Cipro from the material (data not shown). Therefore, only detergent washing was employed, which may have resulted in higher non-specific binding to the controls.

Cipro-dyed C-EDA segments with covalently immobilized ^{125}I-BSA had comparable anti-microbial activity to Cipro-dyed C-EDA segments that were not exposed to protein and Cipro-dyed C-EDA segments that had

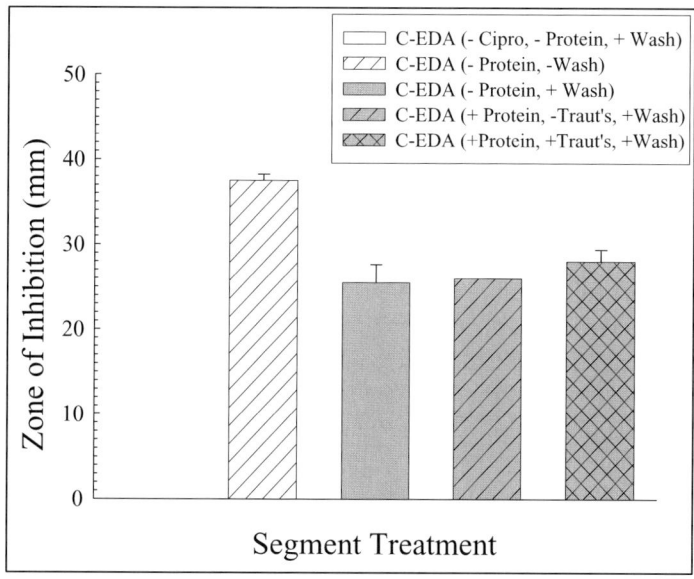

Figure 7.10. Anti-microbial activity of polyester with bound protein

non-specifically bound ^{125}I-BSA (Figure 7.10). Untreated C-EDA segments had no anti-microbial activity. Anti-microbial activity from the unwashed Cipro-dyed C-EDA was greater than these three Cipro-dyed C-EDA segment treatments, which were exposed to various solutions throughout the experimental procedure. Thus, although Cipro concentration is decreased with the various solution incubations, protein attachment does not further decrease anti-microbial activity. The anti-microbial activity of the Cipro-dyed C-EDA segments with covalently bound ^{125}I-BSA is still significant.

Untreated C-EDA, unwashed Cipro-dyed C-EDA, Cipro-dyed C-EDA segments that underwent all solution processes without protein exposure, Cipro-dyed C-EDA segments with non-specifically bound ^{125}I-BSA, and Cipro-dyed C-EDA segments with covalently bound ^{125}I-BSA were evaluated for anti-microbial activity against *Staphylococcus epidermidis* and were thawed at 37°C for 1 h. Control and test segments were embedded into streaked Trypticase Soy Agar (TSA) plates and incubated overnight. Standard 5 μg Cipro Sensi-Discs were used as a positive control. The zone of inhibition each piece was determined, taking the average of three individual diameter measurements.

7.7.5 Potential uses for various modified polyester materials

A majority of the previous and current biomaterial modifications continue to focus on a "magic bullet" approach for biomaterial healing. This unidirectional approach has yet to produce a clinically acceptable bioactive material, potentially due to the simplistic approach taken for such a complex phenomenon. The native tissue that these materials are implanted into possesses a multitude of functions that undergo various responses upon injury such as controlling thrombus formation and orchestrating controlled cellular regrowth. Therefore, the next generation of biomaterials must also possess several characteristics in order to better mimic some of the key functions inherent to native tissue. These individual functions, when incorporated into a single biomaterial surface, will act synergistically, resulting in a novel biomaterial with localized biologic activity to stimulate complete healing of the biomaterial.

Two essential areas will need to be addressed when designing a novel biomaterial: material design and surface biologic characteristics. Material design will permit simulation of the physical characteristics of the native tissue such as compliance and durability. The material must also be easy to handle, as well as to implant (suturability). For the base material, polyester could still be utilized due to the biodurability of the fiber and the potential for creating numerous variations in the material design (i.e., knitting, weaving). Additionally, polyester can be chemically modified, creating functional groups within the polymer structure [132–134]. For these reasons, polyester would be the ideal candidate for the base material due to its proven clinical history as well as the various knitting procedures that are available.

Creation of a polyester material that possesses specific biological properties directly at the material surface can be created *via* immobilization of various proteins. For example, prevention of surface thrombus formation could involve covalent linkage of the potent anti-thrombin agent rHir. Thrombin is a pivotal enzyme in the blood coagulation cascade that is primarily responsible for cleavage of fibrinogen to fibrin [135]. During clot lysis, enzymatically active thrombin is released rendering the vessel susceptible to prompt rethrombosis [136, 137]. Even within a clot, thrombin functions as a smooth muscle cell mitogen [138] are chemotactic for monocytes and neutrophils [139, 140] and an aggregator of lymphocytes. Thus, thrombin that goes unregulated within a clot or pseudointima may lead to cellular infiltration and uncontrolled smooth muscle cell proliferation. rHir is the most potent direct inhibitor of thrombin [141], inhibiting the enzymatic, chemotactic, and mitogenic properties of thrombin [142, 143]. Additionally, rHir has also been shown to inhibit clotbound thrombin [144]. Thus, immobilization of rHir provides an attractive strategy to controlling surface thrombus formation prior to cellular healing.

Another example is a surface that propagates hemostasis and stimulates wound healing is a desired property for hernia repair mesh as well as wound-dressing materials. Activation of the coagulation cascade in order to promote thrombus formation in an effort to control excess bleeding at the injury site is the first step. Thrombin is a pivotal enzyme in the blood coagulation cascade that is primarily responsible for cleavage of fibrinogen to fibrin [145]. During clot lysis, enzymatically active thrombin is released rendering the vessel susceptible to prompt rethrombosis [146, 147]. Even within a clot, thrombin functions as a smooth muscle cell mitogen [138] are chemotactic for monocytes and neutrophil [139, 140] and an aggregator of lymphocytes. Thus, immobilization of thrombin to hernia repair mesh or a wound dressing could expedite hemostasis by directing enzymatic fibrinogen cleavage at the biomaterial surface/injury interface as well as by activating additional clotting factors within the wound (e.g., platelets).

For cellular attachment and proliferation, VEGF, a 42 kDa homodimeric glycoprotein, has been shown to be a potent endothelial cell mitogen and vasopermeability factor [148]. VEGF, which is produced by many different cell types both in tissue culture and *in vivo*, binds to plasma membrane receptors on endothelial cells only with an extracellular transmembrane glycoprotein linked to an intracellular tyrosine kinase domain [149]. This mitogen has also been implicated as a necessary component of wound healing [150, 151]. VEGF in these studies was being released from either a protein scaffold or a viral vector. Thus, covalent linkage of VEGF or another growth factor such as basic fibroblast growth factor to a vascular graft, hernia repair mesh, or a wound dressing would promote initial cellular growth followed by complete healing at the site of injury.

7.8 Conclusion

Polyester is a widely used and useful material. Its usefulness extends into the medical field, where its strength is maintained in implanted devices and materials: it is biodurable. In both medical and non-medical use, it has properties that are less than desirable. A large volume of research has been aimed at modifying the fiber, to make it more dyeable, less soil-retentive, more comfortable, less flammable, and so on. The modifications have been achieved with a wide range of techniques of several different types. These techniques, both in principle and practice, can be used to overcome the limitations that polyester faces in medical use, principally its clotting behavior, lack of infection resistance, and incorporation into body tissue.

References

1. Mohapatra, C. R. Oligomer problem in polyester dyeing and how to reduce it. *Text. Dyer Printer* **1984**, *21*(Nov), 21–25.
2. McIntyre, J. E. Polyester fibres. In: Lewin, M.; Pearce, E. M. (Eds.) *Handbook of Fibre Chemistry*, 2nd edn. Marcel Dekker Inc., New York, **1998**, 1–69.
3. Stevens, K. A.; Brown, P. J. The Characterization and Production of Poly (Ethylene Terephthalate) Polymers and Fibers from Titanium Catalysts. *AATCC Review* **2005**, *5*(3), 17–20.
4. Cook, J. G. *Handbook of Textile Fibres, II Man-Made Fibers*, 5th edn. Merrow, Durham, UK, **1984**.

5. Sawan, S. *Infection-Resistant Materials for Medical Devices and Products*. Technomic Publishing, Lancaster, PA, **1999**.

6. Pourdeyhimi, B. Vascular Grafts: Textile Structures and Their Performance. *Text. Prog.* **1986**, *15*(3).

7. Bide, M.; Phaneuf, M.; LoGerfo, F.; Quist, W.; Szycher, M. In: Edwards, J. V.; Vigo, T. L. (Eds.) *Bioactive Fibers and Polymers*, Chapter 9. ACS Symposium Series 792. American Chemical Society, Washington DC, **2001**.

8. Lewin, M.; Sello, B. (Eds.). *Handbook of Fiber Science and Technology, V2 pt. B Chemical Processing of Fibers and Fabrics,—Functional Finishes*. Marcel Dekker, New York, **1984**.

9. Burkinshaw, S. M. *The Chemistry of Synthetic Fibre Dyeing*. Blackie, Glasgow, **1995**.

10. Schiraldi, D. A.; Martin D. L. A combinatorial method for developing deep dye polyesters. *Text. Res. J.* **2002**, *72*, 153.

11. Masei, Y.; Kato, Y.; Fukui, N. U.S. Patent 3,719,727 (Toyobo).

12. Anon. Fifth International Symposium on Manmade Fibers, **1990**, *4*, 17.

13. Salvio, G.; Stibal, W. Terital MAP—a new antipilling polyester fiber. *Melliand International*, **2002**, *8*(4), 232.

14. Anon. *Adv. Text. Technol.* **2001**, *2*, 1.

15. Foley, F. K. Canadian Patent 919,693. Hercules Inc., **1973**.

16. Dickason, W. C.; Van Sickle, D. E.; McIntire, J. M. U. S. Patent 3,755,498. Eastman Kodak, **1973**.

17. Anon. Abstracts of papers. *Am. Chem. Soc.* National Meeting, **1976**, *172*, 92.

18. Jackson, W. J. Jr.; Darnell, W. R. U. S. Patent 3763644. Eastman Kodak Co., **1973**.

19. Yanagi, M.; Nakamura, I.; Ohosugi, M.; Takizawa, K.; Kasinami, M; Sano, C. U. S. Patent 3658634. Toray Industries, **1972**.

20. Nahta, R. Internal Antistats. *Am. Dyest. Rep.* **1975**, *64*, 41.

21. Anon. Unitika: new conductive polyester fiber. *Chem. Fibers Int.* **2003**, *53*(1), 7, 1p.

22. Kissa, E. Mechanisms of soil release. *Text. Res. J.* **1981**, *51*, 508.

23. Harbereder, P. B.; Bereck, A. Softeners in textile processing. Part 2: Silicone softeners. *Rev. Prog. Color.* **2002**, *32*, 125.

24. Deshpande, A. *Asian Text. J.* **2002**, *11*(10), 64.

25. Nassl, W.; Schreiber, L.; Dirschi, F. Neue Effekte Mit Fluorchemikalien. *Melliand Textilberichte—Int. Text. Rep.* (German Edition), **2002**, *83*(4), 243.

26. Audenaert, F.; Lens, H.; Rolly, D.; Vander, E. P. Fluorochemical Textile Repellents: Synthesis and Applications: A 3M Perspective. *J. Text. Inst.* **1999**, *90*(3), 76.

27. Sherman, P. O.; Smith, S.; Johannessen, B. Textile characteristics affecting the release of soil during laundering. *Text. Res. J.* **1969**, *39*, 449.

28. Gargarine, D. M. *Text. Chem. Colorist* **1978**, *10*, 247.

29. Billie, H.; Eckell, A.; Schmidt, G. Finishing for durable press and soil release. *Text. Chem. Colorist* **1969**, *1*, 600.

30. Inagaki, N. (1996). Plasma-graft copolymerization. In: *Plasma Surface Modification and Plasma Polymerization*. Technomic Publishing Company, Inc., Pennsylvania.

31. Uchida, E.; Uyama, Y.; Ikada, Y. Antistatic properties of surface-modified polyester fabrics. *Text. Res. J.* **1991**, *61*(8), 483–488.

32. Praschak, D.; Bahners, T.; Bossmann, A.; Schollmeyer, E. Melliand Textilberichte. *Int. Text. Rep.* (German Edition) **1997**, *78*(7/8), 531+, 2 pp.; E120, 1 p.

33. Dadestan, M.; Mirzadeh, H.; Sharifi-Sanjani, N. Surface modification of polyethylene terephthalate film by CO_2 laser-induced graft copolymerization of acrylamide. *J. Appl. Polym. Sci.* **2000**, *76*, 401–407.

34. Stannet, V.; Walsh, W. K.; Bittencourt, E.; Liepins, R.; Surles, J. R. Chemical Modification of Fibers and Fabrics with High-Energy Radiation. *J. Appl. Polym. Sci. Appl. Polym. Symp.* **1977**, *31*, 201.

35. Higgins, E. B. Finishing for water repellency. *Text. Inst. Ind.* **1966**, *4*, 255. LSRep 38.

36. Valko, E. I.; Tesoro, G. C. Polyamine resins for the finishing of hydrophobic fibers. *Text. Res. J.* **1959**, *29*, 21.

37. Sello, S. B. Functional Finishes for Natural and Synthetic Fibers. *J. Appl. Polym. Sci. Appl. Polym. Symp.* **1977**, *31*, 229.

38. Kissa, E.; Dettre, R. Subject: Sorption of Surfactants in Polyester Fibers. *Text. Res. J.* **1975**, *45*, 773.

39. Garrett, D. A.; Hartley, F. N. New finishes for Terylene and Terylene blend fabrics. *J. Soc. Dyers Colour* **1966**, *82*, 252.

40. Ferguson, C. A. Hydrophilic finishes for polyester: durability and processing advantages. *Am. Dyest. Rep.* **1982**, *71*, 43.

41. Zeronian, S. H.; Collins, M. J. Surface Modification of Polyester by Alkaline Treatments. *Text. Prog.* **1989**, *20*(2) (Textile Institute), 1–34.

42. Dave, J.; Kumar, R.; Shrivastave, H. C. Study on modification of polyester fabrics: I: alkaline hydrolysis. *J. Appl. Polym. Sci.* **1987**, *33*, 455.

43. Latta, B. Improved tactile and Sorption properties of polyester fabrics through caustic treatment. *Text. Res. J.* **1984**, 766.

44. Yoon, M.-Y.; Kellis, J.; Poulose, A. J. Enzymatic modification of polyester. *AATCC Rev.* **2002**, *2*(6), 33.

45. Gorchakova, V. M.; Izmailov, B. A.; Gartsueva, O. A. *Fibre Chem.* **1999**, *31*(4), 315+, 3 pp.

46. Zahn, H.; Pfeifer, H. Aminololysis of polyethylene terephthalate. *Polymer*, **1963**, *4*, 429–432.

47. Farrow G.; Ravens A. S.; Ward I. M. The degradation of poly(ethylene terephthalate) by methylamine: a study by x-ray and infrared methods. *Polymer* **1962**, *3*, 17.

48. Ellison, M. S.; Fisher, L. D.; Alger, K. W.; Zeronian, S. H. Physical Properties of Polyester fibers degraded by aminolysis and by alkaline hydrolysis. *J. Appl. Polym. Sci.* **1982**, *27*, 247, 257.

49. Avny, Y.; Rebenfeld, L. Chemical modification of polyester fiber surfaces by amination reactions with multifunctional amines. *J. Appl. Polym. Sci.* **1986**, *32*, 4009.

50. Bide, M.; Zhong, T.; Ukponmwan, J.; Phaneuf, M.; Quist, W.; LoGerfo, F. *Proceedings of the Annual International Conference and Exhibition of the American Association of Textile Chemists and Colorists*, Charlotte, NC, October 1–4, 2002. American Association of Textile Chemists and Colorists, PO Box 2215, Research Triangle Park, NC, Oct 1–4, **2002**, 206.

51. Nunn, D. M. (Ed.). *The Dyeing of Synthetic Polymer and Acetate Fibres.* Dyers Company Publications Trust, London, **1979**.

52. Chidambaram, D.; Venkatraj, R.; Manisankar, P. Tensile behavior of polyester yarns modified by solvent-acid mixture pretreatment process. *Indian J. Fibre Text. Res.* **2002**, *27*(2), 199–210.

53. Glasser, W. G.; Wright, R. S. Steam-assisted biomass fractionation. II. Fractionation behaviour of various biomass resources. *Biomass Bioenergy* **1998**, *14*(3), 219–235.

54. Focher, B.; Marzetti, A.; Beltrame, P. L.; Avella, A. Steam exploded biomass for the preparation of conventional and advanced biopolymer-based materials. *Biomass Bioenergy* **1998**, *14*(3), 187–194.

55. Focher, B., et al. Sulphonated aspen pulps. I. Effect of vapour phase cooking temperature on the fiber structure. *Cellulose Chem. Technol.* **1994**, *28*, 629–648.

56. McGonigle, G. Ph.D. Thesis, Modification and characterisation of PET fabric. The University of Leeds, UK, **2001**.

57. Ueda, M.; Tokino, S. Physico-chemical modifications of fibres and their effect on coloration and finishing. *Rev. Prog. Color.* **1996**, *26*, 9–19.

58. Zhang, D.; Sun, Q.; Wadsworth, L. C. Mechanism of corona treatment on polyolefin films. *Polym. Eng. Sci.* **1998**, *38*(6), 965–971.

59. Ryu, J.; Wakida, T.; Takagishi, T. Effect of corona discharge on the surface of wool and its application to printing. *Text. Res. J.* **1991**, *61*(10), 595–601.

60. Song, Q. I.; Netravaili, A. N. Excimer laser surface modification of ultra-high-strength polyethylene fibers for enhanced adhesion with epoxy resins. Part 1. Effect of laser operating parameters. *J. Adhes. Sci. Technol.* **1998**, *12*(9), 957–982.

61. Knittel, D.; Schollmeyer, E. Surface structuring of synthetic fibres by UV laser irradiation. Part III. Surface functionality changes resulting from excimer-laser irradiation. *Polym. Int.* **1998**, *45*, 103–109.

62. Wong, W.; Chan, K.; Yeung, K. W.; Lau, K. S. Chemical modification of poly (ethylene terephthalate) induced by laser treatment. *Text. Res. J.* **2001**, *71*(2), 117–120.

63. Brown, P. J.; Neill, S. M. *Proceedings of the Annual International Conference and Exhibition of the American Association of Textile Chemists and Colorists*, Charlotte, NC, October 1–4, 2002. American Association of Textile Chemists and Colorists, PO Box 2215, Research Triangle Park, NC, Oct 1–4, **2002**, 270.

64. Strobel, M.; Walzak, M. J.; Hill, J. M.; Lin, A.; Karbashewski, E.; Lyons, C. S. A comparison of gas-phase methods of modifying polymer surfaces. *J. Adhes. Sci. Technol.* **1995**, *9*(3), 365–383.

65. Strobel, M.; Branch, M. C.; Ulsh, M.; Kapaun, R. S.; Kirk, S.; Lyons, C. S. Flame surface modification of polypropylene film. *J. Adhes. Sci. Technol.* **1996**, *10*(6), 515–539.

66. Sarmadi, A. M.; Ying, T. H.; Denes, F. Surface modification of polypropylene fabrics by acrylonitrile cold plasma. *Text. Res. J.* **1993**, *63*(12), 697–705.

67. Tsai, P. P.; Wadsworth, L. C.; Roth, J. R. Surface modification of fabrics using a one-atmosphere glow discharge plasma to improve fabric wettabilty. *Text. Res. J.* **1997**, *67*(5), 359–369.

68. Morosoff, N. An introduction to plasma polymerisation. In: d'Agostino, R. (Ed.) *Plasma Deposition, Treatment and Etching of Polymers.* Academic Press Ltd, London, **1990**, 1–84.

69. Sorli, I.; Petasch, W.; Kegel, B.; Schmid, H.; Liebl, G. Plasma processes Part I: Plasma basics, plasma generation. *Inform. Midem* **1996**, *26*(1), 35–45.

70. Saramadi, M.; Denes, A. R.; Denes, F. Improved dyeing properties of SiCl$_4$ (ST)-plasma treated polyester fibres. *Text. Chem. Color.* **1996**, *28*(6), 17–22.

71. Inagaki, N. Interactions between plasma and polymeric materials. *Plasma Surface Modification and Plasma Polymerization.* Technomic Publishing Company, Inc., Pennsylvania, **1996**, 21–41.

72. Okuno, T.; Yasuda, T.; Yasuda, H. Effect of crystallinity of PET and Nylon 66 fibers on plasma etching and dyeability characteristics. *Text. Res. J.* **1992**, *62*(8), 474–480.

73. Wong, K. K.; Tao, X. M.; Yuen, C. W. M.; Yeung, K. W. Low temperature plasma treatment of linen. *Text. Res. J.* **1999**, *69*(11), 846–855.

74. Wróbel, A. M.; Kryszewski, M.; Rakowski, W.; Okoniewski, M.; Kubacki, Z. Effect of plasma treatment on surface structure and properties of polyester fabric. *Polymer* **1978**, *19*(Aug), 908–912.

75. Wakida, T.; Takeda, K.; Tanaka, I.; Takagishi, T. Free radicals in cellulose fibers treated with low temperature plasma. *Text. Res. J.* **1989**, *59*(1), 49–53.
76. Sarmadi, M.; Denes, A. R.; Denes, F. Improved Dyeing Properties of SiCl₄ (ST)-Plasma Treated Polyester Fabrics. *Text. Chem. Colorist* **1996**, *28*(6), 17+, 6 pp.
77. Negulescu, I.; Kwon, H.; Collier, B. J. Determining fibre content of blended textiles. *Text. Chem. Colorist* **1998**, *30*(6), 21–25.
78. Denes, F.; Hua, Z. Q.; Barrios, E.; Young, R. A.; Evans, J. Influence of RF-cold plasma treatment on the surface-properties of paper. *J. Macromolecular Sci.—Pure Appl. Chem.* **1995**, *A32*(8–9), 1405–1443.
79. Gerenser, L. J. XPS studies of *in situ* plasma-modified polymer surface. *J. Adhes. Sci. Technol.* **1993**, *7*(10), 1019–1040.
80. Sapieha, S.; Cerny, J.; Klemberg-Sapieha, J. E.; Martinu, L. Corona versus low pressure plasma treatment: Effect on surface properties and adhesion of polymers. *J. Adhes.* **1993**, *42*(1–2), 91–102.
81. Liang, H.; Sun, Q.; Hou, X. Surface modification of polypropylene microfibre by plasma-induced vapor grafting with acrylic acid. *Chin. J. Polym. Sci.* **1999**, *17*(3), 221–229.
82. Hsieh, Y.-L.; Wu, M. Residual reactivity for surface grafting of acrylic acid on argon glow-discharged poly(ethylene terephthalate) (PET) films. *J. Appl. Polym. Sci.* **1991**, *43*, 2067–2082.
83. Ghenaim, A., Elachari, A., Louati, M., Caze, C. Surface energy analysis of polyester fibers modified by graft fluorination. *J. Appl. Polym. Sci.* **2000**, *75*, 10–15.
84. Hsieh, Y.-L.; Wu, M. Residual reactivity for surface grafting of acrylic acid on argon glow-discharged poly(ethylene terephthalate) (PET) films. *J. Appl. Polym. Sci.* **1991**, *43*, 2067–2082.
85. Kottke-Marchant, K.; Anderson, J.; Umemura, Y.; Marchant, R. Effect of albumin coating on the *in vitro* blood compatibility of polyester arterial prostheses. *Biomaterials* **1989**, *10*, 147.
86. Coleman, R.; Hirsh, J.; Marder, V.; Salzman, E. *Hemostasis and Thrombosis: Basic Principals and Clinical Practices*, 2nd edn. JB Lippincott Company, Philadelphia, PA, **1987**.
87. Shepard, A.; Gelfand, J.; Callow, A.; O'Donnell Jr., T. Complement activation by synthetic vascular prostheses. *J. Vasc. Surg.* **1984**, *1*, 829.
88. Amid, P. K.; Shulman, A. G.; Lichtenstein, I. L. Selecting synthetic mesh for the repair of groin hernia. *Postgrad. Gen. Surg.* **1992**, *4*(2), 150.
89. Chvapil, M.; Holubec, H.; Chvapil, T. Inert wound dressing is not desirable. *J. Surg. Res.* **1991**, *51*(3), 245.
90. Ezzibdeh, M. Y.; Zallat, A. A.; Al-Oraifi, I.; Egail, S. A.; Al-Dayel, A. K.; Sayed, E. E.; Anz, S. A. Internal jugular vein access for hemodialysis using dual-lumen silicon catheters with polyester cuff. www.kfshrc.edu.sa/annals/211_212/00-170.htm. February 2001.
91. Bide, M. J.; Phaneuf, M. D.; LoGerfo, F. W.; Quist, W. C.; Szycher, M. Arterial grafts as biomedical textiles (Chapter). *Bioactive Fibers and Polymers*, ACS symposium series 792, American Chemical Society, Washington DC, 2001.
92. Bide, M. J.; Ozaki, C. K.; Phaneuf, M. D. W.; Quist, W. C.; LoGerfo, F. W. The Use of Dyeing Technology in Biomedical Applications. *Text. Chem. Colorist* **1993**, *25*(12), 15.
93. Phaneuf, M. D.; Quist, W. C.; Bide, M. J.; LoGerfo, F. Modification of polyethylene terephthalate (Dacron) via denier reduction: effects on material tensile strength, weight and protein binding capabilities. *J. Appl. Biomater.* **1995**, *6*, 289.
94. Phaneuf, M. D.; Berceli, S. A.; Bide, M. J.; Quist, W. C.; LoGerfo, F. Covalent linkage of recombinant hirudin to polyethylene terephthalate (Dacron): creation of a novel antithrombin surface. *Biomaterials* **1997**, *18*, 755.

95. Phaneuf, M. D.; LoGerfo, F. W.; Quist, W. C.; Bide, M. J. Surface modification of polyester vascular grafts: Incorporation of antithrombin and mitogenic properties (Chapter). In: Dumitriu, S. (Ed.) *Polymeric Biomaterials.* Marcel Dekker, New York, **2001**.

96. Rumisek, J.; Wade, C.; Kaplan, K.; Okerberg, C.; Corley, J.; Barry, M.; Clarke, J. The influence of early surface thromboreactivity on long-term arterial graft patency. *Surgery* **1989**, *105*, 654.

97. Bascom, J. Gelatin sealing to prevent blood loss from knitted arterial grafts. *Surgery* **1961**, *50*, 947.

98. Drury, J.; Ashton, T.; Cunningham, J.; Maini, R.; Pollock, J. Experimental and clinical experience with a gelatin impregnated polyester prosthesis. *Ann. Vasc. Surg.* **1987**, *1*(5), 542.

99. Guidoin, R.; Marceau, D.; Couture, J.; Rao, T.; Merhi, Y.; Roy, P.-E.; Faye, D. D. L. Collagen coatings as biologic sealants for textile arterial prostheses. *Biomaterials* **1989**, *10*, 156.

100. Barbucci, R.; Magnani, A. Conformation of human plasma proteins at polymer surfaces: The effectiveness of surface heparinization. *Biomaterials* **1994**, *15*(12), 955.

101. Van Der Lei, B.; Bartels, D. F.; Wildevuur, Ch. R. H. The thrombogenic characteristics of small-caliber polyurethane vascular prostheses after heparin bonding. *Trans. Am. Soc. Artif. Intern. Organs* **1985**, *31*, 107.

102. Ma, X.; Mohammad, S. F.; Kim, S. W. Heparin binding on poly(l-lysine) immobilized surface. *J. Colloid. Interface Sci.* **1991**, *147*, 251.

103. Kishida, A.; Ueno, Y.; Fukudome, N.; Yashima, E.; Maruyama, I.; Akashi, M. Immobilization of human thrombomodulin onto poly(ether urethane urea) for developing antithrombogenic blood-contacting materials. *Biomaterials* **1994**, *15*(10), 848.

104. Grainger, D. W.; Okano, T.; Kim, S. W. Protein adsorption from buffer and plasma onto hydrophilic-hydrophobic poly(ethylene oxide)-polystyrene multiblock copolymers. *J. Colloid. Interface Sci.* **1989**, *132*, 161.

105. Silver, J. H.; Hart, A. P.; Williams, E. C.; Cooper, S. L.; Charef, S.; Labarre, D.; Jozefowicz, M. Anticoagulant effects of sulphonated polyurethanes. *Biomaterials* **1992**, *13*(6), 339.

106. Kottke-Marchant, K.; Anderson, J.; Umemura, Y.; Marchant, R. Effect of albumin coating on the in vitro blood compatibility of polyester arterial prostheses. *Biomaterials* **1989**, *10*, 147.

107. Coleman, R.; Hirsh, J.; Marder, V.; Salzman, E. *Hemostasis and Thrombosis: Basic Principals and Clinical Practices*, 2nd edn. JB Lippincott Company, Philadelphia, PA, **1987**.

108. Shepard, A.; Gelfand, J.; Callow, A.; O'Donnell Jr., T. Complement activation by synthetic vascular prostheses. *J. Vasc. Surg.* **1984**, *1*, 829.

109. Gristina, A. Biomaterial-centered infection: Microbial adhesion versus tissue integration. *Science* **1987**, *237*, 1588.

110. Sugarman, B.; Young, E. J. Infections associated with prosthetic devices: Magnitude of the problem. *Infect. Dis. Clin. North Am.* **1989**, *3*(2), 187.

111. Craver, J. M.; Ottinger, L. W.; Darling, R. C.; Austen, W. G.; Linton, R. R. Hemorrhage and thrombosis as early complications of femoropopliteal bypass grafts: Causes, treatment and prognostic implications. *Surgery* **1973**, *74*(6), 839.

112. Griesler, H. P.; Kim, D. U. Vascular grafting in the management of thrombotic disorders. *Semin. Thromb. Hemostas.* **1989**, *15*(2), 206.

113. Herring, M. B. Endothelial cell seeding. *J. Vasc. Surg.* **1991**, *13*(5), 731.

114. Griesler, H. P. Endothelial cell transplantation onto synthetic vascular grafts: Panacea, poison or placebo. In: Griesler, H. P. (Ed.) *New Biologic and Synthetic Vascular Prostheses.* RG Landes Company, Austin, Texas, **1991**.

115. Zilla, P.; Fasol, R.; Preiss, P.; Kadletz, M.; Deutsch, M.; Schima, H.; Tsangaris, S.; Groscurth, P. Use of fibrin glue as a substrate for in vitro endothelialization of PTFE vascular grafts. *Surgery* **1989**, *105*(4), 515.

116. Pierschbacher, M. D.; Ruoslahti, E. Cell attachment activity of fibronectin can be duplicated by small synthetic fragments of the molecule. *Nature* **1984**, *309*, 30.

117. Kuusela, P. Fibronectin binds to Staph aureus. *Nature* **1978**, *276*, 718.

118. Visser, M. T.; van Bockel, H.; van Muijen, N. P.; van Hinsbergh, V. M. Cells derived from omental fat tissue and used for seeding vascular prostheses are not endothelial in origin. *J. Vasc. Surg.* **1991**, *13*(3), 373.

119. Radomski, J. S.; Jarrell, B. E.; Pratt, K. J.; Williams, S. K. Effects of in vitro aging on human endothelial cell adherence to Polyester vascular graft material. *J. Surg. Res.* **1989**, *47*(2), 173.

120. Rosenman, J. E.; Kempczinski, R. F.; Pearce, W. H.; Silberstein, E. B. Kinetics of endothelial cell seeding. *J. Vasc. Surg.* **1985**, *2*(6), 778.

121. Lindblad, B.; Burkel, W. E.; Wakefield, T. W.; Graham, L. M.; Stanley, J. C. Endothelial cell seeding efficiency onto expanded polytetrafluoroethylene grafts with different coatings. *Acta Chir. Scand.* **1986**, *152*, 653.

122. Massia, S. P.; Hubbell, J. A. Human endothelial cell interactions with surface-coupled adhesion peptides on a nonadhesive glass substrate and two polymeric biomaterials. *J. Biomed. Mater. Res.* **1991**, *25*, 223.

123. Ito, Y.; Liu, S. Q.; Imanishi, Y. Enhancement of cell growth on growth factor-immobilized polymer film. *Biomaterials* **1991**, *12*, 449.

124. Gray, J. L.; Kang, S. S.; Zenni, G. C.; Kim, D. U.; Kim, P. I.; Burgess, W. H.; Drohan, W.; Winkles, J. A.; Haudenschild, C. C.; Greisler, H. P. FGF-1 stimulates ePTFE endothelialization without intimal hyperplasia. *J. Surg. Res.* **1994**, *57*, 596.

125. Lin, H. B.; Sun, W.; Mosher, D. F.; García-Echeverría, C.; Schaufelberger, K.; Lelkes, P. I.; Cooper, S. L. Synthesis, surface, and cell-adhesion properties of polyurethanes containing covalently grafted RGD-peptides. *J. Biomed. Mater. Res.* **1994**, *28*, 329.

126. Park, K. D.; Okano, T.; Nojiri, C.; Kim, S. W. Heparin immobilization onto segmented polyurethane surfaces—Effect on hydrophilic spacers. *J. Biomed. Mater. Res.* **1988**, *22*, 977.

127. Kotteke-Marchant, K.; Anderson, J.; Umemura, Y.; Marchant, R. Effect of albumin coating on the *in vitro* blood compatibility of polyester arterial prostheses. *Biomaterials* **1989**, *10*, 147.

128. Phaneuf, M. D.; Dempsey, D. J.; Bide, M. J.; Szycher, M.; Quist, W. C.; LoGerfo, F. W. Bioengineering of a novel small-diameter polyurethane vascular graft with covalently bound recombinant hirudin. *ASAIO J.* **1998**, *44*, M653.

129. Phaneuf, M. D.; Bide, M. J.; Quist, W. C.; LoGerfo, F. W. Merging of biomedical and textile technologies in order to create infection-resistant prosthetic vascular grafts. In: Sawan, S. P.; Manivannan, G. (Eds.) *Anti-microbial/Anti-infective Materials; Principles, Applications and Devices.* Technomic Publishing, Lancaster, PA, **1999**.

130. Phaneuf, M. D.; Ozaki, C. K.; Bide, M. J.; Quist, W. C.; Alessi, J. M.; Tannenbaum, G. A.; LoGerfo, F. W. Application of the quinolone antibiotic ciprofloxacin to polyester utilizing textile dyeing technology. *J. Biomed. Mater. Res.* **1993**, *27*, 233.

131. Bide, M. J.; Phaneuf, M. D.; Ozaki, C. K.; Alessi, J. M.; Quist, W. C.; LoGerfo, F. W. The use of dyeing technology in biomedical applications. *Text. Chem. Colorists* **1993**, *25*(1), 15.

132. Phaneuf, M. D.; Quist, W. C.; Bide, M. J.; LoGerfo, F. W. Modification of polyethylene terephthalate (polyester) via denier reduction: Effects on material tensile strength, weight, and protein binding capabilities. *J. Appl. Biomater.* **1995**, *6*, 289.

133. Phaneuf, M. D.; Berceli, S. A.; Bide, M. J.; Quist, W. C.; LoGerfo, F. W. Covalent linkage of recombinant hirudin to polyethylene terephthalate (polyester): Creation of a novel antithrombin surface. *Biomaterials* **1997**, *18*(10), 755.

134. Phaneuf, M. D.; Bide, M. J.; Dempsey, D. J.; LoGerfo, F. W.; Quist, W. C. Novel bifunctionalized polyester biomaterial surfaces. Proceedings, Surfaces in Biomaterials, Tampa, FL, Apr **2002**.

135. Fenton, J. W. Regulation of thrombin generation and functions. *Semin. Thromb. Hemost.* **1988**, *14*, 234.

136. Gulba, D.; Barthels, M.; Westhoff-Bleck, M.; Jost, S.; Rafflenbeul, W.; Daniel, W. G.; Hecker, H.; Lichtlen, P. R. Increased thrombin levels during thrombolytic therapy in acute myocardial infarction. *Circulation* **1991**, *83*(3), 937.

137. Anderson, H.; Willerson, J. Thrombolysis in acute myocardial infarction. *N. Engl. J. Med.* **1993**, *329*(10), 703.

138. Walz, D. A.; Anderson, G. F.; Fenton, J. W. Responses of aortic smooth muscle to thrombin and thrombin analogues. *Ann. N. Y. Acad. Sci.* **1986**, *485*, 323.

139. Bar-Shavit, R.; Kahn, A.; Wilner, G.; Fenton, J. Monocyte chemotaxis: Stimulation by specific exosite region in thrombin. *Science* **1983**, *220*, 728.

140. Bizios, R.; Lai, L.; Fenton, J.; Malik, A. Thrombin-induced chemotaxis and aggregation of neutrophils. *J. Cell Physiol.* **1986**, *128*(3), 485.

141. Markwardt, F. Pharmacology of hirudin: One hundred years after the first report of the anticoagulant agent in medicinal leeches. *Biochim. Acta* **1985**, *44*, 1007.

142. Obberghen-Schilling, E. V.; Perez-Rodriguez, R.; Pouyssegur, J. Hirudin, a probe to analyze the growth-promoting activity of thrombin in fibroblasts; reevaluation of the temporal action of competence factors. *Biochem. Biophys. Res. Commun.* **1982**, *106*, 79.

143. Fenton, J. W.; Bing, D. H. Thrombin active-site regions. *Semin. Thromb. Hemost.* **1986**, *12*(3), 200.

144. Weitz, J.; Hudoba, M.; Massel, D.; Maganore, J.; Hirsh, J. Clot-bound thrombin is protected from inhibition by heparin–antithrombin III but is susceptible to inactivation by antithrombin III-independent inhibitors. *J. Clin. Invest.* **1990**, *86*, 385.

145. Fenton, J. W. Regulation of thrombin generation and functions. *Semin. Thromb. Hemost.* **1988**, *14*, 234.

146. Gulba, D.; Barthels, M.; Westhoff-Bleck, M.; Jost, S.; Rafflenbeul, W.; Daniel, W. G.; Hecker, H.; Lichtlen, P. R. Increased thrombin levels during thrombolytic therapy in acute myocardial infarction. *Circulation* **1991**, *83*(3), 937.

147. Anderson, H.; Willerson, J. Thrombolysis in acute myocardial infarction. *N. Engl. J. Med.* **1993**, *329*(10), 703.

148. Ferrara, N.; Houck, K. A.; Jakeman, L. B.; Leung, D. W. Molecular and biologic properties of the vascular endothelial growth factor family of proteins. *Endocr. Rev.* **1992**, *13*, 18.

149. Leung, D. W.; Cachianes, G.; Kuang, W. J.; Goeddel, D. V.; Ferrara, N. Vascular endothelial growth factor is a secreted angiogenic mitogen. *Science* **1989**, *246*, 1306.

150. Falanga, V.; Isaacs, C.; Paquette, D.; Downing, G.; Kouttab, N.; Butmare, J.; Badiavas, E.; Hardin-Young, J. Wounding of bioengineered skin: Cellular and molecular aspects after injury. *J. Invest. Dermatol.* **2002**, *19*(3), 653.

151. Kim, B. S.; Chen, J.; Weinstein, T.; Noiri, E.; Goligorsky, M. S. VEGF expression in hypoxia and hyperglycemia: Reciprocal effect on branching angiogenesis in epithelial–endothelial co-cultures. *J. Am. Soc. Nephrol.* **2002**, *13*(8), 2027.

Chapter 8

BIOLOGICAL ACTIVITY OF OXIDIZED POLYSACCHARIDES

Ioan I. Negulescu[1] and Constantin V. Uglea[2]

[1]*Louisiana State University and LSU AgCenter, Baton Rouge, LA 70803, U.S.A.*
[2]*Department of Biomedical Engineering, University of Medicine and Pharmacy, Iasi 6600, Romania*

8.1 Introduction

Among the major classes of biomolecules, carbohydrates allow almost unlimited structural variations. The molecular diversity of carbohydrates offers a valuable tool for drug discovery in the areas of biologically important oligosaccharides, glycoconjugates, and molecular scaffolds by investigating their structural and functional impact. The high density of functional groups per unit mass and the choice of stereochemical linkages at the anomeric carbon have always challenged synthetic chemists toward a multitude of approaches to this rich class of compounds.

The last decade of the past century witnessed the transformation of glycoscience in a worldwide domain. The reason may be that the synthesis of a saccharide chain with biological functions does not involve nucleic acids directly. In other words, in contrast to protein synthesis by genetic information, saccharide synthesis is assisted by a number of enzymes (e.g., glycosidases and glycosyl transferases) acting on a particular position of a molecule on a particular site and at a particular time. Therefore the structure of saccharide chains depends on the environment and often it is uncertain.

Many natural polysaccharides participate in a variety of *in vivo* biochemical reactions. However, it is quite difficult to elucidate the mechanism of their biological activity because of the complex geometry and chemical structure of natural saccharide chains, as well as due to impurities difficult to be removed. In order to understand better the relationship between the biological

J. V. Edwards et al. (eds.), Modified Fibers with Medical and Specialty Applications, 125–143.

activity and the chemical structure of saccharide chains, chemical synthesis of polysaccharides has been attempted [1]. Several types of stereoregular polysaccharides, such as amino and deoxy polysaccharides having a linear or a branched structure were synthesized to this aim by ring-opening polymerization of anhydrosugar derivatives.

There has been an intensified effort in recent years in identifying the biological functions of polysaccharides as related to potential biomedical applications. A large palette of polysaccharides derived from plants, lichens, and algae has been tested. Polysaccharides appear in many different forms in plants. They might be neutral polymers or they might be polyanionic consisting of only one type of monosaccharide, or they might have two or more, up to six different monosaccharide types. They can be linear or branched and they might be substituted with different types of organic groups, such as methyl and acetyl groups. More often than not the biologically active polysaccharides are charged, for example, when the polymers contain uronic acid units, e.g., D-galacturonic acid as in pectic polysaccharides. Other types of polysaccharides isolated from plants used in the traditional medicine were identified as having their biologically active sites in the complementary system, the case of arabinans and arabinogalactans [2].

Similar types of polysaccharides have also been shown to have dissimilar biological activities [3, 4]. Glucans have for a long time been known to have an effect on the immune system. The polymers are normally β-1,3-glucans with a certain degree of branching through C_6. These glucans have been investigated especially in Asian countries, and when it was found that the glucan isolated from the edible mushroom *Lentimus edodes* exhibited a marked anti-tumor effect, the interest in this type of compounds arose [5–7]. In a series of works started early in 1970s Maeda et al. [8–10] reported that lentinan, a polysaccharide composed mostly of β-1,3 and β-1,6 glucosidic linkages, inhibited the growth of Sarcoma-180 transplanted subcutaneously into mice and that the anti-tumor activity was due to a host mediated reaction with participation of thymus or thymus dependent cells (T cells). The authors found that three kinds of protein components deferring from properdin increased markedly in mouse serum soon after lentinum administration. Thus, it appeared that there was a close relationship between the increase of protein components and the anti-tumor activity of polysaccharides. In addition, this fact indicated that at least in an early stage, serum factors could play an important role in the tumor regression, as well as in the cell mediated immune response [10]. A more recent publication on lentinan and related polysaccharides gives a good overview on the studies performed on these types of polymers [11].

Later on Sasaki and Nitta [12] found that the administration of curdlan with an average degree of polymerization DP = 450, at a dose of 5–50 mg/kg for 10 days, had a marked inhibitory effect on subcutaneously (SC) implanted Sarcoma-180. This glucan was also highly inhibitory active at doses of 50 and 100 mg/kg when administrated intraperitoneal (IP). The effect of this polysaccharide is thought to be host mediated because of a lack of *in vitro* activity. In a subsequent work Marikawa and Mizumo [13] indicated that the presence of calcium ions is essential for the anti-tumor activity of curdlan. A similar biologically active polysaccharide has been reported by Bao et al. [14]. The authors isolated a linear $(1\rightarrow3)$-β-D-glucan from the spores of *Ganoderma lucidum* which affected lymphocyte proliferation and had potent stimulating effects on the immunological system.

Another source of biologically active polysaccharide is *Mahonia aquifolium* (Pursh) Nutt. This plant from the Berberidaceae family has been used for a long time in homeotherapy as an organotropic drug for treatment of inflammatory, scaling dermatoses; it is also known as a topical anti-psoriatic drug [15]. The active principle of the mahonia tincture were thought to be the extracted alkaloids [16, 17], but some experiments led to the suggestion that there are probably other components positively influencing the immune mechanisms of human leucocytes. Kardosova et al. [18] proved that the true active component contained by *Mahonia* aqueous–ethanolic extract is a neutral polysaccharide with a linear $(1\rightarrow4)$-β-D-glucan structure. The same authors [19] found that the water-extractable polysaccharide complex from the aerial parts of *Rudbeckia fulgida* var. *sullivantii*, possessed a high anti-tussive activity and Bukovsky et al. [19] reported on significant immunostimulating activity of aqueous–ethanolic extracts from the roots of the same *Rudbeckia* species. In view of these findings, Kardosova and Matulova [19] provided results on isolation and structure identification of the main neutral component of the water-extractable polysaccharide mixture, a fructofuran of the inulin type. Due to its urinary tract tropism, insulin might find wider application in medicine as a drug carrier, especially of drugs to cure urogenital diseases [20].

A new type of polysaccharide was isolated from the Icelandic lichen, *Thamnalia subuliformis*. This polysaccharide, *Thamnolan*, has an unusual structure as it is basically composed of $(1\rightarrow3)$—linked galactofuranosyl units with branched on C_6, and rhamnosyl units being predominantly $(1\rightarrow2)$ linked with branches on C_3 and C_4, while some units are $(1\rightarrow3)$ linked. Xylose is only present as terminal units, while glucose or mannose and galactofuranosyl also are found as terminal units. Glucose and mannose are also $(1\rightarrow4)$ linked. The immunostimulating activity was tested in an *in vitro* phagocytosis assay and anti-complementary assay, and proved to be active in both tests [21].

Interest in the sialic acids has rapidly increased in recent years, especially since their involvement in the regulation of a great variety of biological phenomena was recognized. Based on such observation, it was recognized that sialic acids play a strong, protective role in living cells and organisms. This remarkable function of sialic acids appears to be mainly due to their peripheral position in glucoconjugates and, correspondingly to their frequent, external location in cell membranes [22].

Derivatized and modified polysaccharides such as water-soluble cellulosic derivatives or oxidized carboxymethylcellulose (CMC) exhibit various biological activities, such as oxygen affinity [23]. Blood anti-coagulant activity was certified for sulfonated polysaccharides [1, 24a]. At the same time, it was found that sulfonated polysaccharides have potent anti-human immunodeficiency virus activity (anti-HIV), which increases with the proportion of the amino sugar and branched units on the main chain, but decreases with the increase of deoxy sugar units [1].

Polymers with negative charge on the chain can also function as drugs. Both natural polyanions such as heparin and hyaluronic acid [24b], and synthetic polyanions such as polyacrylic acid and maleic anhydride copolymers exhibit anti-viral, anti-tumor, and anti-fungal activities [25].

The polymer systems explored for pharmaceutical applications include [26–28]:

- polymer drugs, i.e., polymers or copolymers that are themselves physiologically active;
- drug-carrying polymers, i.e., polymers that have active drug bound covalently to a parent polymer backbone;
- time-release drug polymers, i.e., polymers used for sustained or controlled release of drug; the drug may be covered with water-soluble polymer coatings dissolving at different rates and releasing the drug at various times, or the drug may be imbedded into a polymer matrix which diffuses it at specific rates;
- drug-cyclic oligomers inclusion complexes, where the system is established only through non-covalent interactions between the drug and the cyclic carrier;
- site-specific drugs, i.e., polymeric systems that have special chemical groups attached to the polymer carrying the drug; these groups can combine with specific receptors sites on proteins or lipids on the cell surface to achieve specific binding and drug delivery.

Even if macromolecular drugs cause certain therapeutical complications (related to toxicity, reduced capacity of access, etc.) they represent, nevertheless, an alternative to classical drugs.

One of the most interesting possibilities of chemotherapy's development is the utilization of oligomers as drug carriers [29–31]. It maintains the advantages of polymers employed as drug carrier matrices and eliminates some of the disadvantages created by them.

Saponines constitute bioactive substances of vegetal origin, having a wide range of biological activities, such as growing stimulants for various vegetal cultures, as well as anti-oxidant and anti-fungal properties [32]. The chemical structure of such compounds may be viewed as consisting of two distinct regions, namely the oligosaccharide chain and the genine part having a rigid steroid skeleton. A typical saponine structure is given in formula I, were R represents a short normal or branched oligosaccharides chain.

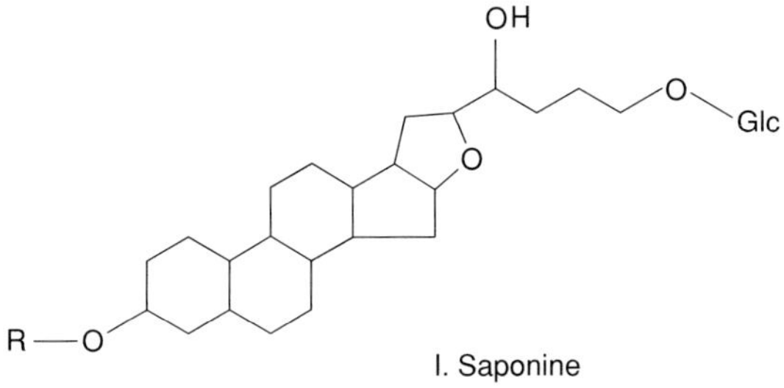

I. Saponine

On the other hand, saponines may be considered as functionalized oligosaccharides, in which the polysaccharide chain forms the hydrophilic part, while the genine part (also known as aglycon) is the hydrophobic part.

In a series of studies, Kintya and Bobeyko [33, 34] put into evidence the possible utilization of such substances as contraceptives, anti-viral, and anti-tumor agents. As to the anti-tumor activity, *in vivo* experiments evidenced that—by means of experimental tumors of the AK 755 (adenocarcinoma of the mammary gland) and C 37 (Sarcoma type)—saponines anti-tumor activity ranges between 6% and 69% (expressed as average tumor retention—ATR), being conditioned by the presence of at least six anhydroglucan units in the chemical structure.

In 1990, Bobeyko et al. [35] utilizing the alcoholic extraction of tomato seeds (*Lycopersicum aesculentum*) obtained Tomatoside® the chemical structure of which is given in formula II.

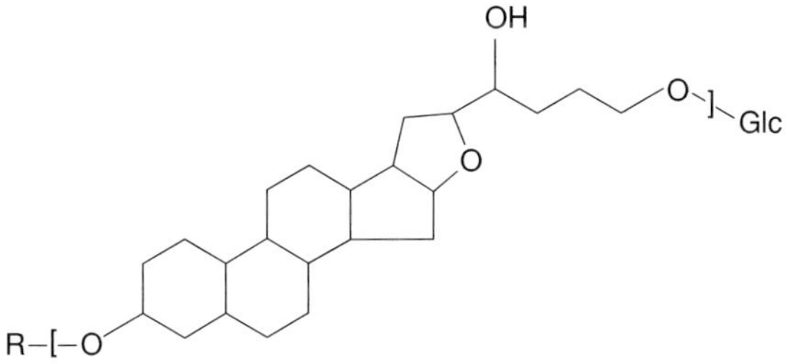

Where:
  [aglycone] is the rest between parentheses
  Glc = rest glucose
  and R is as depicted below

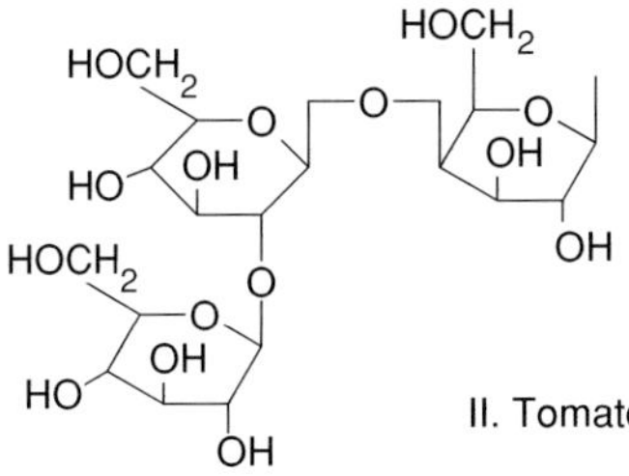

II. Tomatoside, T

Alcoholic extraction of lady's glove (*Digitalis purpurea*) produced a related compound known as Pavstim® (formula III).

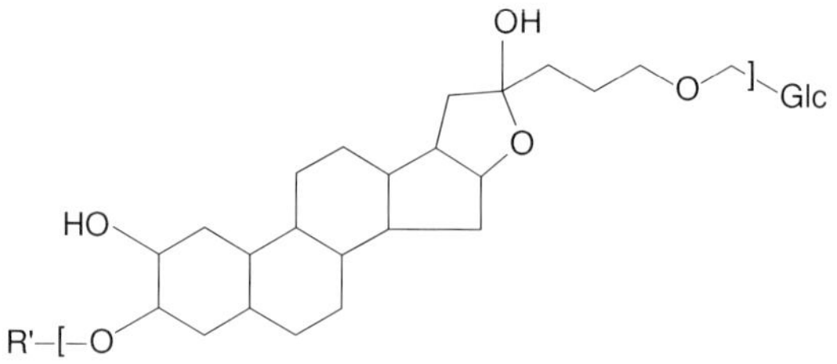

Where:
      [pavstim aglycone] is the rest between parentheses
      Glc = rest glucose
      and R' is as depicted below

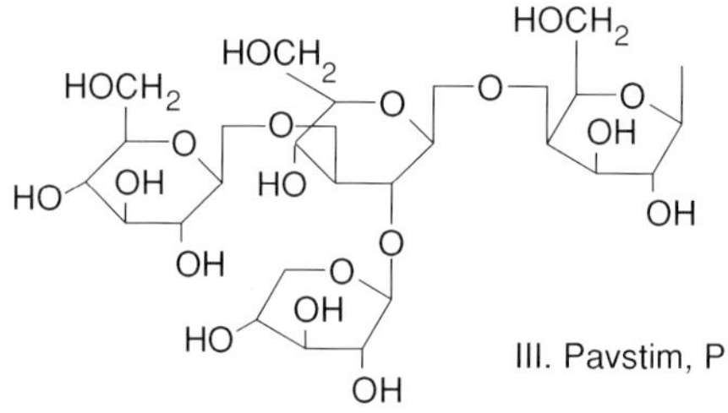

III. Pavstim, P

Viewed globally—at present—the evolution of chemotherapy goes over an asymmetrical period, being deeply anchored on tradition, and quite opaque to any innovation. Consequently, it was proved as incapable of applying the new ideas put forth as early as the beginning of the century, preferred the so called "variation on the same theme", and rejected the perspectives and strategies based on innovative principles.

The science and application of macromolecular drugs may represent a new stage in chemotherapy, quite similar with what earlier decades of the past century represented for physics. Unfortunately, pharmacologists, physicians and— to a considerable extent—biologists remained quite indifferent to such novelties and made no attempts of integrating the polymer science within the fascinating field of the living matter [36].

In our opinion, the interaction between the living matter and the active biological compounds—be they either natural or synthetic—represents a mutual exchange of information. In such a situation, classical drugs are quite primitive and are incapable of reacting promptly against various agents attacking the living organism. The present chapter describes the biological activity of some artificial and natural polyanionic polymers. The data analyzed in the following have been obtained in our laboratories in the investigations initiated as early as 1990.

8.2 Experimental

8.2.1 Oxidation reactions

Commercial cellulose for column chromatography or CMC, both as powders of 100–200 meshes, where treated with sodium metaperiodate at ambient temperature in the dark. The amount of metaperiodate added was 1.4-fold as much as that required by the reaction. The oxidation was performed in a reaction vessel for more than 150 h or at least until the periodate concentration leveled at a constant value; the amount of periodate consumed in the reaction process was determined spectrophotometrically at 290 nm. The excess periodate was decomposed by adding ethylene glycol in the system. The oxidized products were separated by centrifugation into a water-insoluble precipitate and a supernatant liquid. The precipitate was recovered as a colorless powder after a sequential washing with water, 50% ethanol, and ethanol 99.9% and drying in vacuum. In this manner 2,3-dialdehyde cellulose (DAC) or 2,3-dialdehyde carboxymethylcellulose (DACMC) were obtained, DAC and DACMC were used for the synthesis of 2,3-dicarboxycellulose (DCC), formula IV,

IV. 2,3-dicarboxycellulose, DCC

or 2,3-dicarboxy-CMC (DCCMC), formula V. Aqueous suspension (300 mL) containing DAC (15 g) or DACMC (15 g) were further oxidized to this aim with sodium chlorite and acetic acid as described elsewhere [36–39].

V. 2,3-dicarboxy CMC, DCCMC

Tomatoside and Pavstim were obtained by alcoholic extraction as mentioned before [35]. Two-step oxidation with IO_4^- and ClO_2^- of Tomatoside and Pavstim generated polycarboxylic products VI

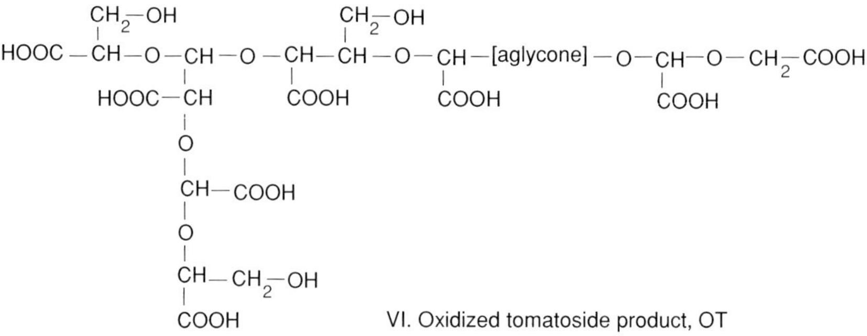

$$HOOC-\underset{\underset{\underset{O}{|}}{\underset{\text{HOOC}-\text{CH}}{|}}}{\overset{\overset{\text{CH}_2-\text{OH}}{|}}{\text{CH}}}-O-\text{CH}-O-\text{CH}-\underset{\underset{\text{COOH}}{|}}{\overset{\overset{\text{CH}_2-\text{OH}}{|}}{\text{CH}}}-O-\underset{\underset{\text{COOH}}{|}}{\text{CH}}-[\text{aglycone}]-O-\underset{\underset{\text{COOH}}{|}}{\text{CH}}-O-\text{CH}_2-\text{COOH}$$

CH—COOH
|
O
|
CH—CH₂—OH
|
COOH

VI. Oxidized tomatoside product, OT

and VII, respectively [40].

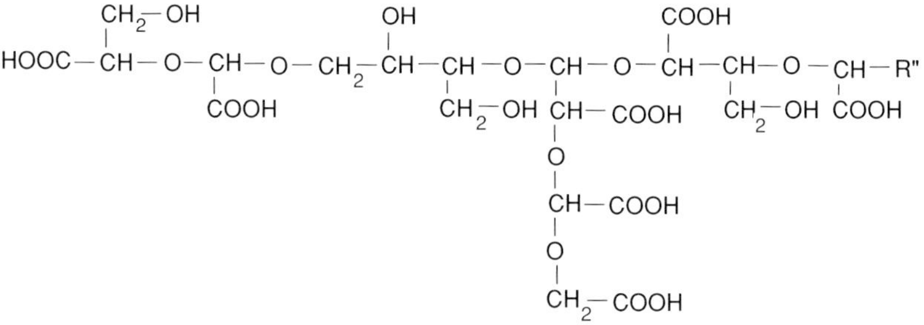

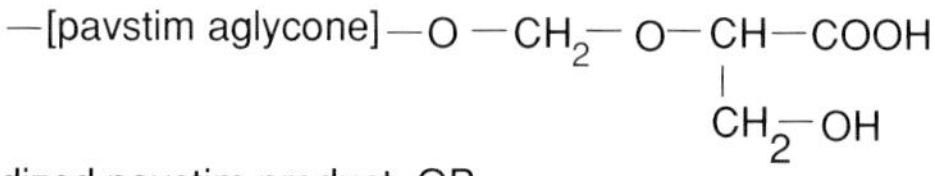

VII. Oxidized pavstim product, OP

Where R" has the structure shown below:

−[pavstim aglycone]−O−CH₂−O−CH−COOH
                                |
                              CH₂−OH

8.2.2 Modification of cellulose polymers

CMC, DCC, and DCCMC were modified with benzocaine (ethyl-*p*-amino benzoate) and *N*-hydroxy-3,4-dihydroxy benzamide (DIDOX) according to literature indications [41–45]. The modified polymers (VIII–XII) were characterized by elemental analysis and FTIR.

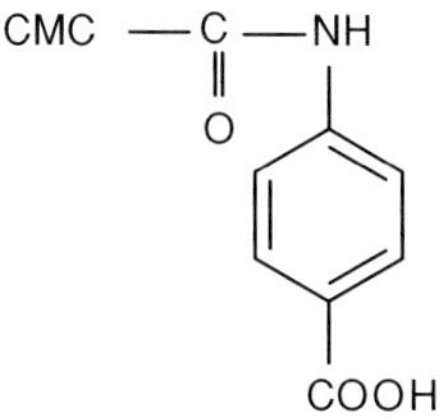

CMC — C — NH
        ||
        O

COOH

VIII. Benzocaine modified CMC, BMCMC

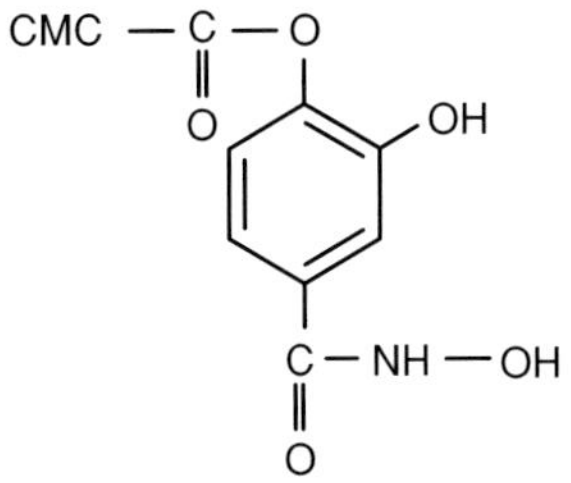

IX. Didox modified CMC, DMCMC

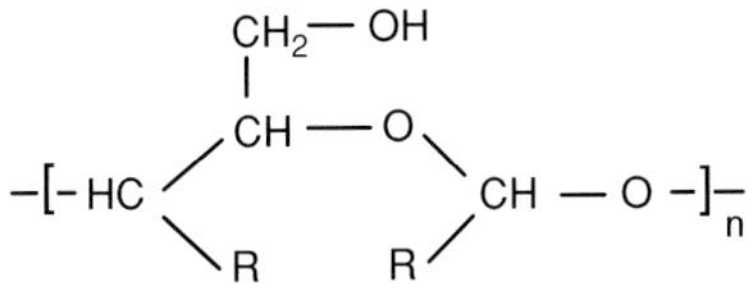

X. Benzocaine modified DCC, BMDCC

where R is:

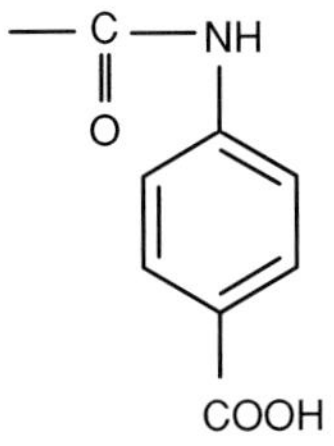

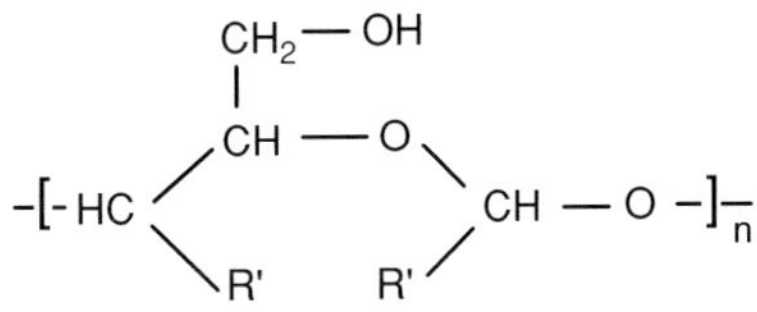

XI. Didox modified DCC, DMDCC

and

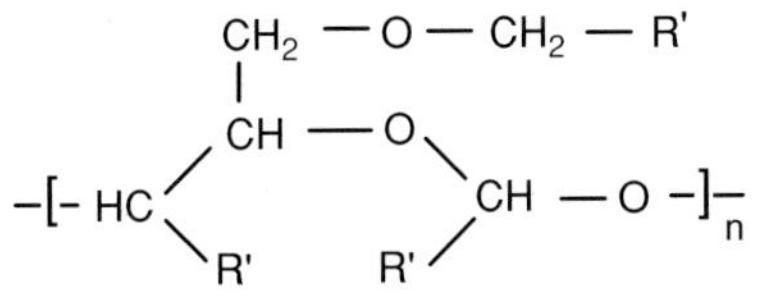

XII. Didox modified DCCMC, DMDCCMC

where R′ is:

8.2.3 Biological evaluation

The anti-viral and anti-tumoral activity of the compounds DCC, DCCMC, BMCMC, DMCMC, BMDCC, DMDCC, DMDCCMC, oxidized Tomatoside, and oxidized Pavstim was studied on groups of 30 (in the case of anti-viral activity) from the Hygiene and Public Health Institute, Virussology Department, and the anti-tumor activity was evaluated with male adult Wistar 10 rats in each group, 11 week old, free from chronic disease obtained from the National Cancer Institute, Bucharest, Romania.

Anti-viral activity. Each group, except the control, received 0.4 mL of COCXACKIE A4 virus in sterile saline solution by sub-occipital insection. Then the mice were placed to the cages and maintained on food and water *ad libitum*. One hour after the virus administration, the mice were injected with 8 mg/kg of the above compounds.

Anti-tumor activity of the synthesized compounds was evaluated with male adult Wistar rats, 11 weeks old, free from chronic diseases.

The animals were treated daily, for 9 days. The tumor bearing rats were divided into experimental groups (10 rats per group). Experimental Okker solid tumors were obtained by subcutaneous injection of cell suspensions, according to Pollak's modification of the published procedures [45, 46], to afford for greater reproducibility and easier routine screening. Starting one day after tumor inoculation, each group received 1.5 mg/day/rat of synthesized compounds. After sacrificing, the standard methods were used for the evaluation of hematological, serological, anatomomorphological parameters and biodistribution. The concentration of the compound in the liver, spleen, and tumor of the animals was determined by liquid chromatography using the method proposed by Markaverich et al. [47] and a Beckman 166 liquid chromatograph with a ODS column (250/4.6 mm) and a mobile phase composed of water–methanol–acetic acid.

8.3 Results and discussion

8.3.1 Anti-viral activity

The COCXACKIE A4 virus caused simultaneous generalized paralysis followed by death within the first 24 h after infection. In the presence of the new

Table 8.1. Comparative anti-viral activity

Compound	Death rate expressed as number of dead mice per day			
	24 h	35 h	55 h	96 h
Control	30	—	—	—
IUT	30	—	—	—
DCC	24	2	3	1
DCCMC	23	1	3	3
BMCMC	20	2	4	2
DMCMC	25	2	3	—
BMDCC	21	3	3	3
DMDCC	25	1	2	3
DMDCCMC	25	1	4	—
T	30	—	—	—
OT	22	3	3	2
P	30			
OP	23	3	4	—

Note: IUT, inoculated and untreated group.

compounds paralysis appeared to begin more gradually and started from the rear part of the animal. Based on the data given in Table 8.1, some preliminary conclusions can be drawn with the respect of the structural–anti-viral activity relationship. The presence of carboxyl groups appear to be essential for the anti-viral activity. The anti-viral effect is enhanced by the number of carboxyl groups per structural unit and by the insertion of benzocaine as a spacer unit. The intensity of the anti-viral effect seems to depend on the presence of lipophilic moieties in the investigated compounds.

8.3.2 Anti-tumor activity

Hematological data indicated that, compared to the control sample group, the tumor presence and the mode of treatment did not significantly change hematological parameters or the serum protein composition (unlisted data). At the same time, the tumor and the treatment mode did not affect visibly the weight changes or the liver weight of rats (unlisted data).

Shown in Table 8.2 is the development of the tumor under the treatment with the synthesized compounds. It appears that the number and the distribution of the carboxyl groups are not the only factors determining the tumor inhibiting effect. Based on these data, the cytotoxicity and the mechanism of the action of studied compounds cannot be fully explained.

On the other hand, some literature data [48–53] evidenced a permanent serious concern for organic compounds with hydroxamic groups, among which, special attention should be given to DIDOX. Unfortunately, in spite of it strong

Table 8.2. Comparative anti-tumor activity

Compound	ATR (%)
DCC	30
DCCMC	35
BMCMC	55
DMCMC	60
BMDCC	60
DMDCC	55
DMDCCMC	60
T	69
OT	10
P	44
OP	17

Note: ATR, average tumoral regression.

cytotoxic effect DIDOX exhibited no suitable selectivity. The synthesis of DIDOX cellulosic and CMC derivatives aimed to a suitable selectivity and diminution of secondary effects.

The results listed in Table 8.2 showed that oxidized and derivatized polysaccharides had a significant anti-tumoral activity. However, the oxidized saponine samples exhibited a lower anti-tumoral activity than the unmodified compounds.

The average tumoral regression, ATR, was determined as follows:

$$\text{ATR} = 100(M_\text{C} - M_\text{T})/M_\text{C} \tag{1}$$

where M_C and M_T represent respectively the average tumoral weight of the control group and the average tumoral weight of the treated groups.

Regarding the biodistribution of investigated compounds, the variation of the concentration ranged between 20 and 100 µg in blood and liver and between 150 and 200 µg in tumoral tissue. These data have been reported earlier [51] and they correlate with the results on solid tumors published by Maeda [54, 55].

Therefore, after the process of extravasation, the polymeric drugs meet the tumor mass. How is it possible for polymeric drug to penetrate into the interstitial tissue of the tumor from the blood compartment? To obtain an answer to this question, we should have a proper understanding of the unique feature of the tumor vasculatures.

It is now well accepted that, because of the activation of the kinin-generating cascade and the secretion of vascular permeability factor, blood capillaries at tumor tissues develop in a considerably high density with enhanced permeability due to the loose of the interendothelial junctions. This process leads to enhanced passive transport of macromolecular substances, such as proteins and macromolecular drugs, across the blood vessel into the interstitial space of the tumorous tissues. On the other hand, the development of lymphatic drainage system is insufficient in tumorous tissues, resulting in a poor tissue drainage of macromolecular substances. Consequently, macromolecular substances may exhibit a considerable accumulation in the tumor due to the synergistic effect of the increased vascular permeability and the decreased tissue drainage. This effect has been studied by Maeda and is termed enhanced permeability and retention (EPR) [54, 55]. It has now become one of the major guiding principles in drug targeting using polymeric carriers. Macromolecular accumulation due to the enhanced permeability of blood vessels is also expected to take at the site of inflammatory reactions, as a result of several causes, including microbial infections [55–58].

8.3.3 Mechanism of action

Polymers have many limitations when used either as drugs or as drug carriers [59]. One of the most serious limitations is the existence of cell membrane barrier. Polymer normally cannot enter cells by diffusion through a membrane or by active process *via* membrane proteins. As early as the 1980s it was reported that the normal mechanism whereby such a polymer passes the cell membrane is by endocytosis [60]. This represents engulfment by an infolding of the plasma membrane with formation of a cytoplasmic vacuole (phagosome). Following uptake, fusion with the enzyme-containing lysosome yields the digestive vacuole. The endocytosis of polymers is initiated by their adsorption on the cell uptake of large particles and is influenced by various factors such as molecular weight, chemical structure, and lipophilicity.

In the case of oral administration, the polymer drugs, prior to the endocytosis, must be involved in a transcytosis, process by which it is assimilated in blood circulation. In transcytosis [61], a polymer or oligomer can be shuttled across a cellular barrier by first binding to a small piece of the plasma membrane and subsequently becoming entrapped inside as a small vesicle. The polymer or oligomer containing vesicles can be processed within the cell, translocated across the cell cytosol, and then are able to release their content *via* fusion with the plasma membrane on the opposite side of the cell monolayer.

Both the above-mentioned processes are enhanced or inhibited by structural characteristics of the various polymeric drugs such as chemical structure, molecular weight and its distribution, lipophilicity, and hydrophilicity [62, 63].

The mechanism of anti-viral activity of polyanionic polymers has not been completely defined and a general structure–activity relationship in regard to this effect still needs to be performed. The major models of anti-viral action that have been considered are

a) direct inactivation of virus;
b) inhibition of virus replication;
c) interferon induction;
d) stimulation of phagocytosis and inflammation;
e) specific immunoenhancement of humoral or cell mediated immune responses against the virus;
f) enhancement of macrophage anti-viral functions.

Polyanions can directly inactivate viruses [64]. However, the levels required are greater than those needed for anti-viral activity *in vivo*.

Inhibition of virus replication by mechanisms similar to conventional chemotherapic drugs also does not appear to play an important part in polyanionic resistance *in vivo* [65].

Considerable efforts has been directed toward determining whether induction of the anti-viral protein interferon accounts for anti-viral action of polyanions [66]. There is no clear evidence that systemic induction of interferon play any role in the anti-viral of polycarboxylic polysaccharides [67].

Because polyanions are immunoregulators, attention has been directed toward specific immunoenhancement as the mode of anti-viral action. Specific immunostimulation, however, does not appear to play a proeminent role in the anti-viral activity of polyanions and especially of the polycarboxylic polysaccharides [68].

Investigations that demonstrated no correlation between interferon induction and anti-viral activity provided the first suggestion that activated macrophages might be involved in the anti-viral action of polyanions. Our data substantiate this hypothesis [34]. Moreover, modified CMC-activated peritoneal macrophages exhibit potent anti-viral activity *in vitro* [34]. Thus we have demonstrated that activated macrophages have the capacity for anti-viral activity; however, proofs are still needed that these cells are the major mode of action *in vivo* in animals treated with polyanions.

A possible mechanism for the activity of the synthesized polymers on tumor growth may be related to coupling of the polymers to tumor antigen. However, the action of polyanions on a wide range of enzymes, alteration of the isoelectric point of protein, and anti-viral action all indicate alternative concepts of anti-tumor action.

The anti-tumor activity of polyanions is apparently not due to direct cytotoxicity [69]. However, our observations [70] that a certain synthetic polyanions was selectively cytotoxic for HeLa tumor cells supported the hypothesis that polyanion's anti-tumor activity is due to the direct cytotoxicity and may be also mediated by activated macrophages. Additional experiments will be required to determine polyanion's role in the process of tumor inhibition and the mechanism(s) by which activated macrophages are selectively cytotoxic for tumor cells.

At the present time the basis for this discrimination is unknown but it must be assumed that the polycarboxylic polymers or the activated macrophages can recognize a feature of tumor cell that is inconsistent with that found on a normal cell. For some time new researchers have been trying to delineate singular differences between normal and tumor cells in a given system. Tumor cells have been found to exhibit differences in lecithin binding and agglutination, membrane microviscosity, enzyme activity, and cytochemical structure when compared to normal cells. However, the mechanism by which polyanionic or activated macrophages recognize tumor cells remains to be elucidated.

Conclusions

Significant anti-viral and anti-tumor activity of polycarboxylic polymers was achieved by derivatizing cellulose and CMC with benzocaine and DIDOX. The chemical structure of the synthesized compounds was confirmed by analytical methods. The *in vivo* evaluation of the polymers provided some qualitative insights with respect to the mechanism of the polymer transport through the cell membrane.

Quo vadis pharmaceutical chemistry?

We liken the problem of designing a drug to that of finding a street address in Tokyo. As any visitor to Japan may know many areas of that old city have no street signs and no house numbers. But if one knew the right neighborhood, and could knock on every door in that neighborhood, the problem would be solved. Structural techniques and high-throughput screening offer the hope that we can start our search for new drugs in the right neighborhood. Libraries of natural products, synthetic oligonucleotides and combinatorial small organic compounds give us the capability to knock on literally thousands of doors. Our designs no longer have to be perfect, or even nearly perfect, the first time. This is the recipe for a revolution.

So, believe we really do stand on the threshold of a new era in pharmaceutical chemistry. To summarize, here is what I think the pattern will be ever the next 20 years or so. Target identification will come primarily from genomics and basic cell biology research aided by natural products that define new pathways and molecules to be inhibited or activated. If there is any structural information about some molecules—usually natural products—that are already known to bind to that target, we will make small directed libraries of compounds that will contain one or more high nanomolar leads. If all we have is an assay, we will be able to convert it to a high-throughput screen and fish out several hits from large diversity libraries and pools of natural products. Optimization of these hits into leads will be produced by medicinal chemistry, abetted where necessary by directed combinatorial methods of making analyses and high-resolution structures of lend compound/target complexes.

Development from leads to drugs will follow much the same strategy. Clinical trials will be conducted with several compounds in parallel, with the power performers dropping out until the best drug emerges at the end even though we still might not quite be able to make "designer" drugs. We believe that these new tools for drug discovery—all of which derive from basic research, be the way—will enable us to find leads and develop them into drugs two to three times faster than that has been possible. We always could call spirits from the very deep. This time, they just might come.

References

1. Yoshida, T. Synthesis of polysaccharides having specific biological activities. *Prog. Polym. Sci.* **2001**, *26*, 379–441.
2. Yamada, H.; Kyohara, H. In: Wagner, H. (Ed.) *Immunomodulatory Agents from Plants.* Birkhauser Verlag, Basel, **1999**, 161.
3. Stimple, M.; Prolsch, A.; Wagner, H. Macrophage activation and induction of macrophage cytotoxicity by purified polysaccharide fractions from the plant *Echinacea purpurea. Infect. Immun.* **1984**, *46*, 845–851.
4. Yamada, H. Pectic polysaccharides from Chinese herbs: structure and biological activity. *Carbohydr. Polym.* **1994**, *25*, 269–275.
5. Chihara, G. Inhibition of mouse sarcoma 180 by polysaccharides from *Lentimus edodes* (berk) sing. *Nature* **1969**, *222*, 687–689.
6. Chihara, G.; Maeda, Y. Y. Fractionation and purification of the polysaccharides with marked antitumour activity, especially lentian, from *Lentimus edodes* (berk) sing an edible mushroom. *Cancer Res.* **1970**, 30, 2776–2779.
7. Sasaki, T. The extracellular polysaccharides of *Rhizobium japonica* compositional studies. *Carbohydr. Res.* **1976**, *47*, 99–101.
8. Maeda, Y. Y.; Chihara, G. Lentinan, a new immuno-accelerator of cell-mediated responses. *Nature* **1971**, *229*, 634–638.
9. Maeda, Y. Y.; Chihara, G. Unique increase of serum proteins and action of antitumor polysaccharides. *Nature* **1974**, *252*, 250–251.
10. Maeda, Y. Y.; Chihara, G. Carboxymethyl pachymaran, a new water soluble polysaccharide with marked antitumor activity. *Nature* **1971**, *233*, 486–488.
11. Maeda, Y. Y. In: Wagner, H. (Ed.) *Immunomodulatory Agents from Plants.* Birkhauser Verlag, Basel, **1999**, 203.
12. Sasaki, T.; Nitta, K. Dependence on chain length of antitumor activity of $(1 \rightarrow 3)$-β-D-glucan from *Alcaligenes faecalis* var. myxogenes, IFO 13140, and its acid degraded products. *Cancer Res.* **1978**, *38*, 379–385.
13. Marikawa, K.; Mizumo, D. Calcium-dependent and -independent tumoricidal activities of PMN leukocytes by a linear β-1,3-D-glucan in mice. *Cancer Res.* **1986**, *46*, 66–72.
14. Bao, X. F.; Duan, J. Y.; Fany, J. N. Chemical modification of the $(1 \rightarrow 3)$-alpha-D-glucan from spores of *Ganoderma lucidum. Carbohydr. Res.* **2001**, *223*, 127–140.
15. Galle, K.; Wagner, H. Analytical and pharmacological studies on *Mahonia aquifolium. Phytomedicine* **1994**, *1*, 59–64.
16. Haensel, R. D. A plant antipsoriatic. *Apoth. Ztg.* **1992**, *132*, 2095–2101.

17. Bezakova, L. Lipoxygenase inhibition and antioxidant properties of bisbenzylisoquinoline alkaloids isolated from *Mahonia aquifolium*. *Pharmazie* **1996**, *51*, 758–762.

18. (a) Kardosova, A.; Alfoldi, J.; Machova, E.; Kostolova, D. Polysaccharides in the antipsoriatic *Mahonia* extract; structure of a (1→4)-β-D-glucan. *Chem. Pap.* **2001**, *55*, 192–195; (b) Kardosova, A. Water-extractable polysaccharide of *Rudbeckia fulgida* posses anti-tissue activity. *Chem. Pap.* **1997**, *51*, 52–57.

19. Kardosova, A. A fructofuran from the roots of *Rudbeckia fulgida*. *Coll. Czech. Chem. Commun.* **1997**, *62*, 1799–1803.

20. Kardosova, A. (4-*O*-methyl-α-D-glucurono)-D-xylan from *Rudbeckia fulgida*, var. *sullivantii*. *Carbohydr. Res.* **1998**, *308*, 99–105.

21. Olafsdottir, E. S. Rhamnopyranosylgalacto-furanan a new immunological active polysaccharide from *Thomanolia subuniforms*. *Phytomedicine* **1999**, *6*, 273–279.

22. Kiefel, J. M.; von Itzstein, M. Recent advances in synthesis of sialic acid derivatives and sialylmimetics as biological probes. *Chem. Rev.* **2002**, *102*, 471–490.

23. Zygmont, D. Effect of polyanionic polymers on haemoglobin–oxygen bonding properties. *Int. J. Biol. Macromol.* **1987**, *9*, 343–345.

24. (a) Negulescu, I.I.; Simionescu, C.I.; Capla, M.; Borsig, E. Heparinization of organic polymeric films. *Memoirs Romanian Acad. Sci. Sections* **1989**, *12*(1), 145–154. (b) Brand, C.; Vert, M.; Petitow, M. Extrasulfation of heparin: effect on chemical structures and anticoagulant activity. *J. Bioact. Compat. Polym.* **1989**, *4*, 269–284.

25. Uglea, C. V.; Panaitescu, L. Synthetic polyaminic macromolecules with antiviral and antitumor activity. *Curr. Trends Polym. Sci.* **1989**, *2*, 241–251.

26. Chien, Y. W. New developments in drug delivery systems. *Med. Res. Rev.* **1990**, *10*, 477–504.

27. Fulten, D. A.; Stoddart, J. F. Neoglycoconjugates based on cyclodextrins. *Bioconjug. Chem.* **2001**, *12*, 655–672.

28. Phillips, M.; Ottenbrite, R.M. Polymers in biological systems. *ACS Symp. Ser.* **1988**, *362*, 123–139.

29. Ferruti, P. New polymeric and oligomeric matrices as drug carriers. *CRC Crit. Rev. Ther. Drug Carrier Syst.* **1986**, *2*, 175–241.

30. Uglea, C. V.; Dumitriu, C. Oligomers as "physical catalysts" of biological processes (Chapter 15). In: Dumitriu, S. (Ed.) *Polymeric Biomaterials*. Marcel Dekker Inc., New York, N.Y., **1994**.

31. Uglea, C. V. *Oligomers Technology and Applications*. Marcel Dekker Inc., N.Y., New York, **2000**.

32. Dimolgo, A. S. Structures-actvity correlations for antioxidant and antifungal properties of steroid glycosides. *Zhur. Bioorg. Chem.* **1985**, *11*, 408–413 (in Russian).

33. Kintya, P. K.; Bobeyko, W. A.; Lopatin, P. B.; Sofina, Z. P. Steroid glycosides. Glycosides of *Rokogenine*. S U 677410, **1978** (in Russian).

34. Kintya, P. K. Natural steroid bioregulators. *Rast. Res. Zhur.* **1988**, *2*, 263–272 (in Russian).

35. Bobeyko, W. A.; Kyntia, P. K.; Danka, I. V. Thermal decompostion of furosanol glycoside-tomatoside. *J. Thermal Anal.* **1990**, *36*, 243–253.

36. Arshady, R. Biodegradable microcapsule drug delivery systems: manufacturing methodology, release control, and targeting prospects. *J. Bioact. Compat. Polym.* **1990**, *5*, 315–319.

37. Besemer, A. C.; De Nooy, A. E. J.; Van Bekkum, H. Methods for the selective oxidation of cellulose: preparation of 2,3-dicarboxycellulose and 6-carboxycellulose. *ACS Symposium Series* **1998**, *688*, 73–82.

38. Hofreiter, B. T. Chlorous acid oxidation of periodate-oxidized starch. *J. Am. Chem. Soc.* **1957**, *79*, 6457–6462.

39. Hofreiter, B. T. Hydrogenolysis of dialdehyde starch to erythritol and ethylene glycol. *Anal. Chem.* **1957**, *27*, 1930–1939.

40. Iurea, D.; Uglea, C. V.; Kyntia, P. K. Modified steroidal glycosides with potential biological activity. In: Waller, R.; Yamasachi, K. (Eds.) *Saponines Used in Traditional and Modern Medicine* (Adv. Exp. Med. Biol.). Plenum Press, New York, NY, **1996**, 111–116.

41. Uglea, C. V.; Apetroaiei A.; Offenberg, H.; Negulescu, I. I. Anesthesine modified polysaccharides. Synthesis and characterization. *Polym. Prep.* **1992**, *33*, 96–97.

42. Uglea, C. V.; Albu, I. N. Drug delivery systems based on inorganic materials: 1. Synthesis and characterization of a zeolite–cyclophosphamide system. *J. Bioact. Compat. Polym.* **1994**, *9*, 448–461.

43. Uglea, C. V.; Ottenbrite, R. M. Polysaccharides as antiviral and antitumor support (Chapter 24). In: Dumitriu, S. (Ed.) *Biomedical Applications of Polysaccharides*. Marcel Dekker, Inc., New York, NY, **1994**.

44. Gavat, C.; Chiruta, R.; Iacob, E.; Uglea, C. V. Anionic polymers with biological activity I. Benzocaine modified CMC. *Rev. Med. Chir.* **2003**, *107* (Suppl. 2), 56–61 (in Romanian).

45. Gavat, C.; Chiruta, R.; Iacob, E.; Uglea, C. V. Anionic polymers with biological activity, II. Oxidized CMC. *Rev. Med. Chir.* **2003**, *107* (Suppl. 2), 62–65 (in Romanian).

46. Pollack, V. A.; Fidler, I. J. Use of young mice for selection of subpopulations of cells with increased metastatic potenial from nonsygeneic neoplasms. *J. Natl. Cancer Inst.* **1982**, *69*, 137–145.

47. Markaverich, B. M.; Gregory, R. R.; Alejandro, M. A. Methyl p-hydroxyphenyl lactate indenification in rat liver extracts. *J. High Resolut. Chromatogr. Chromatogr. Commun.* **1988**, *11*, 605–607.

48. Anonymous. New ribonucleotide reductase inhibitors; Didox and trinidox exhibit antiretrovirus activity in several marine animals models. *AIDS Res. Hum. Retroviruses* **1995**, *1* (Supp. 1), 11.

49. Simmonds, R. J. The geometry of *N*-hydroxy-methyl compounds. *J. Chem. Soc. Perkin Trans.* **1993**, *13*, 1399–1404.

50. Spridon, D.; Panaitescu, L.; Vatajanu, A.; Buruiana, E.; Uglea, C. V. The structure biological activity relation in dihydroxamic acids. *Roum. Biotechnol. Lett.* **1997**, *2*, 131–146.

51. Ghitler, N.; Panaitescu, L.; Uglea, C. V. The evaluation of the concentration of some antitumoral agents in organs and biological fluids by reversed phase HPLC. *Studii Cercet. Stiintifice (Biology)* **1996**, *1*, 91–94 (in Romanian).

52. Tepelus, V.; Panaitescu, L.; Uglea, C. V. Benzamide derivatives with cytotoxic and cytostatic activity. *Studii Cercet. Stiintifice (Biology)* **1996**, *1*, 95–98 (in Romanian).

53. Uglea, C. V.; Spridon, D. Biological activity of hydroxamic compounds. In: Wise, R. (Ed.) *Handbook of Biomaterials*. Marcel Dekker Inc., New York, NY, **2000**, 1011–1021.

54. Matsumura, Y.; Maeda, H. A new concept for macromolecular therapeutics in cancer chemothearopy of turnoritropics accumaltion of proteins and the antitumour agents manics. *Cancer Res.* **1986**, *46*, 6387–6392.

55. Maeda, H.; Seymour, L.; Miyamoto, Y. Conjugates of anticancer agents and polymers: Advantages of macromolecular therapeutic *in vivo*. *Bioconjug. Chem.* **1992**, *3*, 351.

56. Toyocumi, T.; Singhal, A. K. Synthetic carbohydrate vaccines based on tumor-associated antigens. *Chem. Soc. Rev.* **1995**, *24*(4), 231–247.

57. Monsigni, M.; Roche, A. G.; Midoux, P.; Mayer, R. Glycoconjugates as carriers for specific delivery of therapeutic drugs and genes. *Adv. Drug Deliv. Res.* **1994**, *14*, 1–55.

58. Wadhwa, M. B.; Rice, K. G. Receptor mediated targeting. *J. Drug Targeting* **1995**, *3*, 111–135.

59. Duncan, R.; Kopecek, J. Soluble synthetic polymers as potenial drug carriers. *J. Adv. Polym. Sci.* **1984**, *57*, 53–72.
60. Ringsdorf, H. Stucture and properties of pharmacologically active polymers. *J. Polymer Sci. Symp.* **1975**, *68*, 135–147.
61. Wei-Chiang S.; Jiansheng W.; Taub, M. In: Dunn, R. L.; Ottenbrite, R. M. (Eds.) *Polymeric Drug Delivery System (ACS Symp. Ser.)* **1991**, *469*, 117–127.
62. Uglea, C. V.; Negulescu, I. I. Synthesis and Characterization of Synthetic Oligomers. CRC Press, Boca Raton, **1991**, 239–326.
63. Uglea, C. V.; Medvighi, C. Medical applications of synthetic oligomers (Chapter 15). In: Polymeric Biomaterials. Marcel Dekker Inc., New York, NY, **1994**.
64. Mc Cord, R. S.; Morahan, P. S. Antiviral efffect of pyran against systemic infection of mice with *herpes symplex* type 2. *Antimicrob. Agents Chemother.* **1976**, *10*, 28–35.
65. Papas, T. S. Inhibition of RNA-dependent DNA polymerase of avian myeloblastosis virus by pyran copolymer. *Proc. Natl. Acad. Sci.* **1974**, *71*, 367–376.
66. Breslow, D. S. Divinyl ether-maleic anhydride (pyran) copolymer used to demostrate the effect of molecular weight on biological activity. *Nature* **1973**, *246*, 160–169.
67. Billiau, A. Mechanism of antiviral activity in vivo of polycarboxylases which induce interferon production. *Nature* **1971**, *232*, 183–191.
68. Hirsch, M. S. Comparison of spot-blot and microtiterplate method for the detection of HIV-1 PCR products. *J. Immunol.* **1992**, *108*, 1312–1319.
69. Morahan, P. S. Antitumour action of pyran copolymer and tiprone against Lewis lung carcinoma and B16 melanoma. *Cancer Res.* **1974**, *34*, 506–513.
70. Uglea, C. V.; Negulescu, I. I.; Siminescu, C. I.; Offenberg, H.; Grecianu, A. Benzocaine modified maleic anhydride-cyclohexyl-1,3-dioxepin copolymer: Preparation and potenial medical applications. *Polymer (London)* **1993**, *34*, 3298–3301.

Chapter 9

BIOLOGICAL ADHESIVES

José María García Páez and Eduardo Jorge-Herrero

Unidad de Biomateriales, Servicio de Cirugía Experimental, H.U. Clínica Puerta de Hierro, c/San Martín de Porres 2; 28035 Madrid, Spain

9.1 Introduction

The need to effectively manage very important problems like hemostasis and tissue sealing has had a strong influence on the development of modern surgical techniques. A group of chemical products (natural and synthetic materials) known as sealants, glues, or tissue adhesives has been developed to reduce bleeding and promote tissue sealing. For example, it has been shown that the use of a cyanoacrylate adhesive contains the bleeding from gastric varices with a higher rate of success and a lower rate of mortality than the administration of ethanolamine oleate, a sclerosing agent [1]. In blepharoplasty, an eye surgery technique, octyl-2-cyanoacrylate, a cyanoacrylate approved by the U.S. Food and Drug Administration (FDA) has been used with excellent results in terms of quality when compared with the use of sutures [2]. Three types of adhesives have been utilized in cardiovascular surgery: fibrin glues, which are resorbable but do not provide strong adhesion and require rapid healing of the tissue; enbucrilates, which have been used successfully for left ventricular free wall rupture [3] but produce a marked exothermic reaction and are unstable; and biological glues. The latter have been employed to bond pericardial patches and reinforce sutures. In aortic dissection, a very serious clinical situation, a bioadhesive is used to bond the proximal and the distal edges of the dissected aorta, which are then sutured. The mechanical behavior of the bioadhesive in this clinical situation has yet to be characterized and the association of sutures appears to be indispensable [4].

Chemically, tissue glues and adhesives can be defined as any substance with characteristics that allow for polymerization [1]. This chemical polymerization

J. V. Edwards et al. (eds.), Modified Fibers with Medical and Specialty Applications, 145–158.
© 2006 *Springer. Printed in the Netherlands.*

must also hold tissues together. Glues and adhesives are materials whose attachment to a surface principally involves molecular attraction.

Certain logical features are required of surgical glues. First of all, the chemical substance employed as the adhesive must remain present and preserve its chemical characteristics long enough for the tissues to bond well without any additional supports and for the necessary time. Evidently, a rapid degradation of the adhesive before the healing process is completed would be counterproductive. The degradation of the adhesive should always occur after that time. Some authors describe the mechanism of action of a good adhesive as a result of two different physical forces. The glue is usually spread on each of two objects; it is held to them by adhesion involving intermolecular forces between the two dissimilar materials. With an effective glue, adhesion and cohesion are about as strong as the internal cohesion of the objects to be joined.

Finally, it must be safe. The agent should not create more problems than it solves. When first introduced, fibrin sealants were banned from use in the United States because of the risk of the transmission of infections. Not until blood products could be adequately screened for these pathogens were they approved. Safety is a critical issue. Like all biomaterials, adhesives should meet certain safety norms, which are regulated by different international organizations. Adhesives and glues must promote tissue healing without the risk of infection or viral transmission. To obtain these properties, the different manufacturers must test their products at least for acute, subacute, local, and systemic toxicity according to well-established protocols that recommend that the tests be carried out in each individual component, in the final product (if there are two or more components mixed together to obtain the final glue or adhesive) and in its degradation products.

As these materials will act in living organisms, it is necessary to take into account that the human body is a very aggressive environment. It is a saline medium with a temperature of 37 °C, so certain conditions—minimum tissue toxicity at the application sites, parallel sealing and biodegradability times to get a proper healing time, wettability; ease of utilization in the surgical theater, and of course, low cost—must be met.

At the present time, tissue adhesives are being used in a number of surgical specialties, but all of them have to offer the same properties or qualities. They must be easy to use (as mentioned above), have fast action in bleeding systems, undergo no exothermic reaction during polymerization, have sufficient strength for each type of tissue to which they are applied and produce no inflammatory reactions.

We review some of the general characteristics of these adhesives and describe our experience (in experimental models) in the mechanical behavior of biological glues that could potentially be utilized, alone or in combination with sutures, to join inert biological tissues that have been chemically treated. The biological tissue employed was glutaraldehyde-fixed calf pericardium similar to that employed in the manufacture of the valve leaflets of cardiac bioprostheses.

9.2 Fibrin glues

Fibrin glues are employed in surgery to help bond tissues during wound healing and repair. Carless et al. [7] published a systematic review of the randomized controlled trials carried out to study the efficiency of fibrin sealants in 2002. Being resorbable, their use in implants or bioprostheses is prohibited.

9.2.1 Composition

A commercially prepared fibrin sealant comes in a kit that includes two lyophilized components. One component contains a pooled fibrinogen factor XIII concentrate, which is dissolved in an antifibrinolytic solution (aprotinin). The other component is bovine thrombin reconstituted with 40 mM $CaCl_2$. The kit also includes a double-barreled syringe system that releases equal volumes of fibrinogen and thrombin through a needle. Historically, the main problem with fibrin sealants was the high associated risk of hepatitis transmission from the human plasma pooled to obtain fibrinogen. The composition of commercial solutions varies considerably. The fibrinogen concentration ranges between 50 and 115 mg/mL, whereas that of factor XIII is 5–80 IU/mL. The thrombin concentration ranges from 200 to 600 u/mL^2.

9.2.2 Mechanism of action

Fibrin sealants are employed for tissue healing and topical hemostasis, so their action mimics the blood coagulation cascade (Figure 9.1) to obtain a final semirigid, insoluble, fibrin clot by mixing the two components. Thrombin catalyzes the step that transforms fibrinogen into fibrin monomers and, with calcium ions as a cofactor, activates factor XIII. Fibrin then starts to polymerize by means of electrostatic interactions and hydrogen bonds. In the presence of calcium, active factor XIIIa converts non-covalent bonds to covalent bonds, which render the cross-linked structure of a fibrin clot.

The fibrin clot is degraded by physiological fibrinolysis. The antifibrinolytic agent, aprotinin, supplied in the kit slows the breakdown of the clot by plasmin, the fibrinolytic agent of our organism.

To prevent virus transmission, fibrin sealants are subjected to a variety of chemical treatments to ensure a safe product, free from virus. At present, fibrinogen can be obtained from individual units of blood plasma that are previously tested for the associated risk of hepatitis and human immune deficiency virus (HIV). In addition, for viral inactivation, commercial solutions are purified

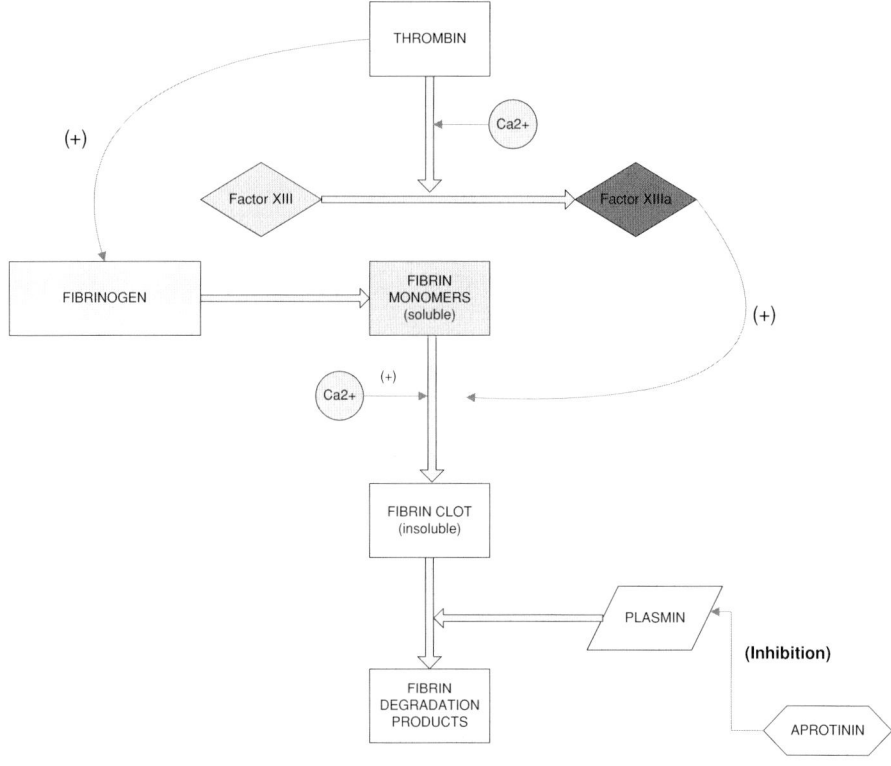

Figure 9.1. Mechanism of clotting factor interactions

using a two-step vapor-heating method at 60 and 80 °C, or other methods such as pasteurization (10 h in an aqueous solution at 60 °C), detergent treatments, nanofiltration, chromatography, ultraviolet C light, etc. A combination of these treatments is preferable because none of them is 100% effective. The methods employed by the manufacturers offer a sufficient margin of safety against HIV and hepatitis B and C viruses. In 2000, Gosalbez et al. reported a parvovirus B19 transmission attributed to the use of fibrin sealant [3], but there are no cases of serious viral transmission reported in the literature.

9.2.3 Mechanical behavior

A study of the behavior of fibrin sealants when employed in combination with conventional sutures showed no improvement in the resistance or elasticity of biological tissue samples. However, a more homogeneous mechanical behavior was observed in sutured samples in which a fibrin glue was subsequently applied to the holes created by the suture. These results do not have a clear explanation and further studies will be required to confirm or refute their validity [10].

Table 9.1. Mean tensile stresses at the time of rupture of the samples

Series	No.	Mean (Mpa)	Standard deviation	95% Confidence interval
PPI	12	6.83	2.34	5.35, 8.32
PI	12	7.06	2.43	5.52, 8.61
PC	12	10.40	2.66	8.72, 12.10
PD	12	8.27	3.18	6.24, 10.29
PPD	12	8.70	2.71	6.98, 10.41

PPI: samples from the left side of the pericardial sac, sutured at a 45° angle and glued. PI: pericardium from the left side, sutured at a 45° angle. PC: control pericardium. PD: pericardium from the right side, sutured at a 90° angle. PPD: pericardium from the right side, sutured at a 90° angle and glued.
PC versus PI: $p = 0.027$; PC versus PPI: $p = 0.016$.

9.2.4 Mechanical characteristics

For the study of the mechanical characteristics of these adhesives, we analyzed a commercially available fibrin glue, Tissucol® Immuno 5.0, that is authorized for medical use. This sealant is manufactured from two components of human origin: a lyophilized protein concentrate to be dissolved in a solution containing aprotinin and lyophilized human thrombin, which is reconstituted with a calcium chloride solution (Baxter AG, Vienna, Austria).

The study consisted of subjecting 60 samples of calf pericardium, a material that is utilized in the manufacture of bioprosthetic valve leaflets, to tensile testing. A running suture from edge to edge, at a 45° (PI) or 90° (PD) angle with respect to the principal axis, using 5.0 Prolene was performed in 24 samples. Another 24 samples (PPI and PPD) were sewn in the same manner, after which the holes made by the suture were reinforced with the fibrin glue. Twelve samples were neither sutured nor glued to serve as a control group (PC). All the samples were then subjected to tensile testing until rupture and the mean tensile stresses being exerted at that point were compared. These stresses, expressed in MPa, are shown in Table 9.1.

In the control series (PC), the mean value was 10.40 MPa. In the sutured and glued series, the mean values ranged between 6.83 and 8.70 MPa. Statistically significant differences were observed only when series PC was compared with the series sutured at a 45° angle (PI) ($p = 0.027$) and that sewn at a 45° angle and glued ($p = 0.016$).

9.2.5 Study of the elongation (strain)

The stress in MPa was analyzed for an elongation, or deformation, of 15%. The findings are shown in Table 9.2. In the control series (PC), the mean value was 3.15 MPa, which was significantly different when compared with the mean value in the other series, which ranged between 0.92 and 1.98 MPa ($p = 0.001$ to $p = 0.0041$).

Table 9.2. Tensile stress in MPa for an elongation of 15%

Series	No.	MPa	Standard deviation	95% Confidence interval
PPI	12	1.81	1.05	1.14, 2.47
PI	12	1.98	0.93	1.38, 2.57
PC	12	3.15	1.29	2.33, 3.97
PD	12	1.55	1.00	0.91, 2.18
PPD	12	0.92	0.53	0.59, 1.26

PC versus PPI: $p = 0.013$; PC versus PI: $p = 0.041$; PC versus PD: $p = 0.002$; PC versus PPD: $p = 0.001$.

The results obtained confirm the loss of resistance to tensile stress in the samples sutured from edge to edge, regardless of the angle of the suture line, when subjected to this 15% elongation.

This does not imply that the sutured and the glued series (PPI and PPD) present greater elasticity, although their behavior is seen to be more uniform, as shown by linear regression analysis comparing the tensile stress in the control series with that of the other series (Table 9.3).

The homogeneity of the samples, and thus of the results, is of the utmost importance for predicting the durability and safety of devices made with bio-materials.

In conclusion, fibrin glues add little to the resistance or elasticity of the samples, although their contribution to the homogeneity of the samples, if confirmed, should be taken into account.

9.3 Biological glues

Gelatin–resorcinol–formaldehyde blue (GRF): This glue (also known as French Blue) was developed in the 1960s by Cooper and Falb [11] as an alternative to the cyanoacrylates. The mixture of gelatin and resorcinol (3:1) and addition of an 18% formaldehyde solution renders a sealant which cross-links tissues (covalently and avidly) in less than 1 min.

Table 9.3. Linear regression coefficients, where PC represents the independent variable (x), and PPI, PI, PD, and PPD are dependent variables (y)

Series	a_1	a_2	r
PPI (y)	0.542	0.018	0.996
PI (y)	0.586	0.031	1.000
PD (y)	0.542	0.018	0.999
PPD (y)	0.542	0.018	0.999

r: correlation coefficient.

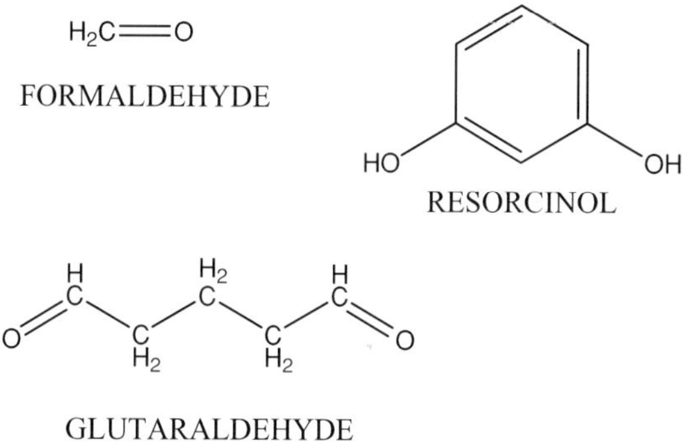

$$H_2C \!=\! O$$

FORMALDEHYDE

RESORCINOL

GLUTARALDEHYDE

Figure 9.2. Chemical structure of the components of some BioGlues

BioGlue: Evolved from GRF glue, it is composed of 10% glutaraldehyde and 45% bovine albumin. When these two components are combined, the aldehyde groups of glutaraldehyde form stable imine cross-links with the amino groups of the proteins (Figure 9.2). The polymerization of the adhesive commences immediately and the reaction is completed in less than 2 min.

Both GRF and BioGlue have been used in a variety of soft tissue applications, but technical problems and the cytotoxicity of formaldehyde and glutaraldehyde have limited their application to acute type A aortic dissection.

The use of and results with these sealants are controversial. Some authors, in a retrospective 20-year analysis, report the use of GRF as extremely useful during initial emergency surgery for acute type A dissection, making the procedure much easier and safer [12]. Others who have reviewed its use report alterations of the aortic wall, unacceptable long-term complications and a high incidence of recurrent aortic regurgitation [4, 12, 13]. Still others prefer full root replacement to treat dissection of the aortic root [15]. These controversial data should be considered with caution. Success with biomaterials of this type (as with other types of surgical procedures or pharmacological treatments) depends on the experience of the surgical team, the amount of the product applied, and the physical conditions of the tissue, which is to be sealed. In other words, the correct use of the glue and previous training on the part of the surgeon (if inexperienced) in *in vivo* models prior to employing it in patients are essential aspects.

9.3.1 Mechanical characteristics

We compared the mechanical behavior of two commercially available adhesives, the biological glue "BioGlue" and the cyanoacrylate Loctite 4011, as bonding agents in 18 samples of a biological tissue, calf pericardium, which were cut in half and then glued with a 1 cm^2 overlap [16]. The major objective

Table 9.4. Results after tensile testing until rupture in samples glued with
Loctite 4011 with BioGlue

Series	No.	Mpa	SD	95% CI	kg	SD	95% CI
Loctite	12	0.15*	0.05	0.12, 0.18	1.56**	0.49	1.25, 1.88
BioGlue	6	0.04*	0.01	0.02, 0.05	0.41**	0.15	0.25, 0.57

SD: standard deviation. 95% CI: 95% confidence interval.
*$p = 0.001$.
**$p = 0.001$.

of this assay was to determine the mechanical resistance to tensile testing of the samples after being rejoined. The mean tensile stresses at rupture are shown in Table 9.4. They are expressed in MPa and machine kg (the load to which the samples were being subjected at rupture according to the tensile testing machine) for a better comparison.

In both cases, the samples repaired with the biological glue showed a statistically significant loss of resistance when compared with the samples bonded with the cyanoacrylate ($p = 0.001$).

Perhaps the greatest problem with the biological glue is that, in order for it to be effective and achieve the necessary resistance, it has to be applied over a dry surface, a difficult condition to meet in this assay in which the calf pericardium was treated with glutaraldehyde at the time of gluing. It would appear to be equally difficult to achieve a dry surgical field for the repair of aortic dissection [4].

9.3.2 Study of the strain (elongation)

To assess the elongation that the samples underwent, we determined the mathematical fit of the values for stress (y) and strain (x) in the two series. The best fit corresponded to a third-order polynomial, the shape of which is expressed as $y = a_1x + a_2x^2 + a_3x^3$. The coefficients of the resulting curves appear in Table 9.5.

Table 9.5. Coefficients of the stress–elongation (strain) curves,
showing the mechanical behaviors of the series glued with Loctite
4011 and with BioGlue

Series	a_1	a_2	a_3	R^2
Loctite	−0.07	4.22	−8.03	0.861
BioGlue	−0.02	0.98	−1.09	0.919

$y = a_1x + a_2x^2 + a_3x^3$, where y is the stress in MPa and x the per unit elongation.
R^2: coefficient of determination.

The analysis of these functions shows that both adhesives allow a marked elongation of the samples, approximately 20%, at very low levels of stress: 0.1 MPa for the tissue glued with the cyanoacrylate and 0.04 MPa for that repaired with the biological glue.

9.4 Cyanoacrylate adhesives

Cyanoacrylate tissue adhesives are pale yellow or transparent liquid monomers that polymerize rapidly through an anionic mechanism in the presence of weak bases such as water, alcohol, or amino groups (from proteins) that come in contact with the tissue surfaces. It is an exothermic reaction that yields a strong flexible film that bonds the wound edges.

Methylcyanoacrylate was the first material tested but was ruled out because of its rapid *in vivo* degradation. Longer-chain alkylcyanoacrylates (Figure 9.3) such as *n*-butyl cyanoacrylate and octylcyanoacrylate were developed to avoid this problem. Butyl cyanoacrylates have poor tensile strength, a circumstance that limits their use to small lacerations and incisions. The development of a longer-chain (octylcyanoacrylate) improved the performance of these biomaterials as adhesives for wound repair.

Cyanoacrylates in contact with living tissues in a moist environment polymerize rapidly to create a thin elastic film of high tensile resistance which

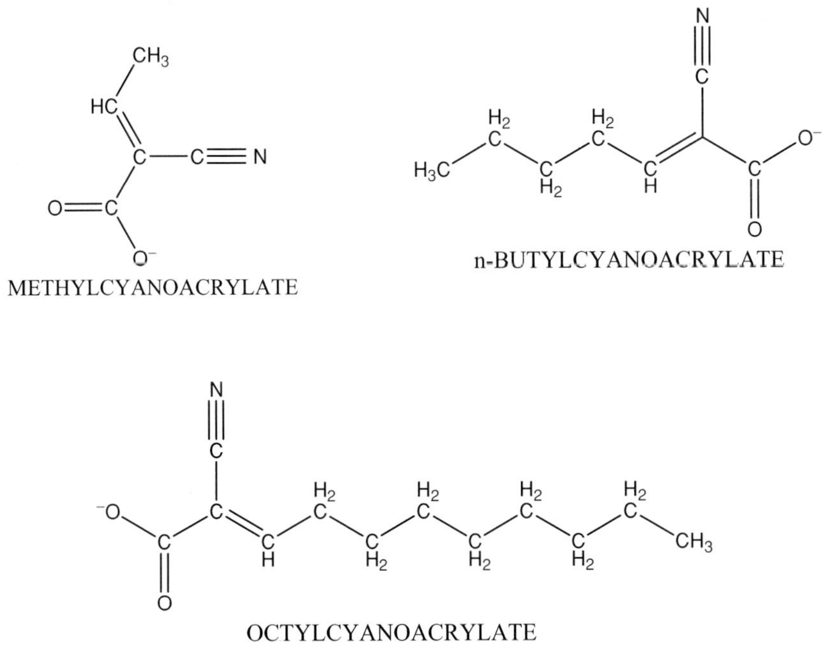

Figure 9.3. Cyanoacrylates used in biological glues

guarantees firm adherence, stronger than that obtained with fibrin glues. Cyanoacrylates provide a flexible, water-resistant coating that is not impaired by blood or organic fluids and that also inhibits microbial growth. Easy to apply, these adhesives do not require the use of local anesthetics. The polymerization time varies as a function of the type of tissue with which the glue comes in contact and the nature of the fluids present. Polymerization starts after 1–2 s and takes about 1–1.5 min. Some authors prefer the superior physical properties of octylcyanoacrylate compared with butyl cyanoacrylate to repair facial lacerations [17]. As advantages, cyanoacrylates slough off spontaneously within 5–10 days, eliminating the need for suture removal.

Thus, it can be said that they are more cost-effective than sutures or staples because of the reduced need for follow-up and practitioner time [17–19].

Some disadvantages of the cyanoacrylates are the decreased tensile strength and the requirement that the areas to be treated be dry. Moreover, the application of an excessive amount of product, in addition to prolonging setting time, increases the exothermal reaction associated with polymerization, with possible thermal damage to the tissue. Thus, the glue should never be applied inside a wound over mucous membranes. It should also be avoided over areas of frequent friction, such as hands or feet, because of the risk of detachment of the adhesive.

Singer and Thode reviewed the literature on octylcyanoacrylates and reported that the current generation of octylcyanoacrylates (high-viscosity formulations) can be used successfully in a wide variety of clinical and surgical settings for multiple types of wounds involving most of the surface of the human body [20]. In an experimental study in rats, using another formulation, ethyl-2-cyanoacrylate, the histopathological analysis of vascular, myocardial, and pulmonary tissue sections demonstrated that there were no significant differences between sutures and ethyl-2-cyanoacrylate in controlling hemorrhage and air leakage [21].

9.4.1 Mechanical characteristics

We assessed the mechanical characteristics and stability over time of a commercially available cyanoacrylate, Glubran 2 (*n*-butyl 2-cyanoacrylate (monomer)/methacryloxysulfolane (monomer) (GEM, s.r.l., Italy). It is a transparent, instantaneous biological adhesive with low viscosity that is authorized for medical use according to the VSP norm (class VI). *(footnote) The results of the study are currently pending publication.

9.4.2 Resistance to rupture

Samples of calf pericardium, cut from the pericardial sac in two perpendicular directions, longitudinally, or from root to apex, or transversely, were subjected to tensile testing.

The pericardium had previously been treated for 24 h with 0.625% glutaraldehyde (pH 7.4) prepared from a commercially available solution of 25% glutaraldehyde (Merck) at a ratio of 1/50 (w/v), in 0.1 M sodium phosphate buffer. Then 60 samples, 30 cut in longitudinal direction (series LP) and 30 cut transversely (series TP), were cut in half and glued with Glubran 2, with an overlap of 0.5 cm (for a total surface area of 1 cm^2). Twelve additional samples, 6 longitudinal (series LC) and 6 transverse (series TC), were used as controls.

The samples were stored until the assay at 4 °C in saline (0.9% NaCl) plus two antibiotics: streptomycin at a concentration of 333 µg/mL and penicillin at a concentration of 2000 U/mL. Six samples each from series LP and TP were subjected to uniaxial tensile testing, always in the direction of the principal axis of the sample, until rupture 7, 30, 60, 90, and 120 days after being glued. The controls were assayed on day 7. The trials were performed on an Instron TTG4 tensile tester (Instron Ltd., High Wycombe, Buck, U.K.), which determines tensile stress and the elongation, or strain, it produces.

The mean results at rupture in kilogram, according to the load being exerted by the machine at that moment, are shown in Tables 9.6 and 9.7. While Table 9.6 shows a statistically significant loss of resistance in the glued samples when compared with the controls on day 7, Table 9.7 demonstrates that the results in the glued samples assayed up to and on day 120 do not change significantly, indicating the stability of the bond.

9.4.3 Elastic behavior

We have observed no loss of elasticity over the 120-day period that could be attributed to a hardening of the tissue secondary to the use of the adhesive. On day 120, the elastic behavior of the glued samples was maintained or even improved, a circumstance that suggests that there are no deleterious effects in this respect over time.

The analysis of the deformation, or elongation, of the samples was of most interest in the series assayed 120 days after being glued (Table 9.8). We observed a reversible deformation after rupture, that is, once the region affected by the

Table 9.6. Comparison of the means in the control series (LC and TC) and the glued series (LP and TP) cut longitudinally and transversely, respectively, on day 7 after gluing

Series	No.	Mean (kg)	Standard deviation	95% Confidence interval	p
LC	6	13.35	2.71	10.51, 15.19	*
TC	6	12.24	3.67	8.38, 16.09	**
LP	6	1.56	0.54	0.99, 2.12	*
TP	6	1.87	0.34	1.50, 2.23	**

*LC versus LP: $p = 0.001$.
**TC versus TP: $p = 0.001$.

Table 9.7. Comparison of the means in the glued series (LP and TP) cut longitudinally and transversely, respectively, on days 30, 60, 90, and 120 after gluing

Series	No.	Mean (kg)	Standard deviation	95% Confidence interval
Day 30				
LP	6	1.28	0.51	0.73, 1.81
TP	6	1.10	0.66	0.41, 1.79
Day 60				
LP	6	1.04	0.23	0.79, 1.29
TP	6	1.18	0.56	0.59, 1.77
Day 90				
LP	6	1.55	0.63	0.88, 2.22
TP	6	1.10	0.89	0.17, 2.03
Day 120				
LP	6	1.67	0.62	1.02, 2.32
TP	6	1.39	0.54	0.83, 1.96

rupture had amply surpassed the elastic limit [22]. To explain this phenomenon, it is necessary to take into account the differences in the elastic behaviors of the various regions of a given sample, with loads concentrated in the area near the glued portion. The collagen fibers of that zone would absorb these loads, undergoing permanent deformation and rupture, while the collagen fibers farther from the repair would be subjected to less stress, not surpassing the elastic limit. Once the load was eliminated, they would recover their original length [23].

Finally, we wish to point out that this method of bonding using cyanoacrylates demonstrates considerable resistance, but probably not sufficient for their use without other bonding elements in bioprostheses or implants. However, it allows a surprising degree of elasticity in tissue samples subjected to tensile testing.

Table 9.8. Percentages of elongation, or deformation, in series cut longitudinally (LP) and transversely (TP) 120 days after being glued

Elongation	No.	Mean (%)	Standard deviation	95% Confidence interval
Irreversible				
LP	6	10.13	7.95	1.78, 18.47
TP	6	8.55	9.49	1.42, 18.51
Reversible				
LP	6	14.51	7.41	6.73, 22.28
TP	6	12.43	5.86	6.27, 18.58
Total				
LP	6	23.89	3.48	20.24, 27.55
TP	6	20.97	4.23	16.53, 25.41

We obtained similar results, with stable resistance to tensile stress for up to 150 days and marked elasticity of the bond, with Loctite 4011, a commercially available ethyl cyanoacrylate described above [24], which is also authorized for medical use according to the VSP norm (class VI).

References

1. Oho, K.; Iwao, T.; Sumino, M.; Toyonaga, A.; Tanikawa, K. Ethanolamine oleate versus butyl cyanoacrylate for bleeding gastric varices: a nonrandomized study. *Endoscopy* **1995**, *27*, 349–354.
2. Greene, D.; Koch, R. J.; Goode, R. L. Efficacy of octyl-2-cyanoacrylate tissue glue in blepharoplasty. A prospective controlled study of wound-healing characteristics. *Arch. Facial Plast. Surg.* **1999**, *1*, 292–296.
3. Gosalbez, J.; Cofiño, J.; Llosa, J.; Naya, J.; Valle, J. The use of biological glue (CGR) in the treatment of the left ventricular ruptures. *Scand. J. Thorac. Cardiovasc. Surg.* **1996**, *30*, 74.
4. Bingley, J. A.; Gardner, M. A.; Stafford, E. G.; Mau, T. K.; Pohlner, P. G.; Tam, R. K.; Jalali, H.; Tesar, P. J.; O'Brien, M. F. Late complications of tissue glues in aortic surgery. *Ann. Thorac. Surg.* **2000**, *69*, 1764–1768.
5. Reece, T. B.; Maxey, T. S.; Kron, I. L. A prospectus on tissue adhesives. *Am. J. Surg.* **2001**, *182*, 40S–44S.
6. Donkerwolcke, M.; Burny, F.; Muster, D. Tissue and bone adhesives—Historical aspects. *Biomaterials* **1998**, *19*, 1461–1466.
7. Carless, P. A.; Anthony, D. M.; Henry, D. A. Systematic review of the use of fibrin sealant to minimize perioperative allogeneic blood transfusion. *Br. J. Surg.* **2002**, *89*, 695–703.
8. Soffer, E.; Ouhayoun, J. P.; Anagnostou, F. Fibrin sealants and platelet preparations in bone and periodontal healing. *Oral Surg. Oral Med. Oral Pathol. Oral Radiol. Endod.* **2003**, *95*, 521–528.
9. Hino, M.; Ishiko, O.; Honda, K. I.; Yamane, T.; Ohta, K.; Takubo, T.; Tatsumi, N. Transmission of symptomatic parvovirus B19 infection by fibrin sealant used during surgery. *Br. J. Haematol.* **2000**, *108*, 194–195.
10. García Páez, J. M.; Jorge-Herrero, E.; Carrera, A.; Millán, I.; Cordón, A.; Rocha, A.; Martín Maestro, M.; Morales, S.; Castillo, J. L. Fijación de mmateriales biolaógicos mediante suturas y adhesivos biológicos. *Biomecánica* **2003**, *11*, 30–38.
11. Cooper, C. W.; Falb, R. D. Surgical adhesives. *Ann. N. Y. Acad. Sci.* **1968**, *146*, 214–224.
12. Bachet, J.; Goudot, B.; Dreyfus, G.; Banfi, C.; Ayle, N. A.; Aota, M.; Brodaty, D.; Dubois, C.; Delentdecker, P.; Guilmet, D. The proper use of glue: a 20-year experience with the GRF glue in acute aortic dissection. *J. Card. Surg.* **1997**, *12*, 243–253.
13. Kirsch, M.; Ginat, M.; Lecerf, L.; Houel, R.; Loisance, D. Aortic wall alterations after use of gelatin–resorcinol–formalin glue. *Ann. Thorac. Surg.* **2002**, *73*, 642–644.
14. von Oppell, U. O.; Karani, Z.; Brooks, A.; Brink, J. Dissected aortic sinuses repaired with gelatin–resorcinol–formaldehyde (GRF) glue are not stable on follow up. *J. Heart Valve Dis.* **2002**, *11*, 249–257.
15. Kazui, T.; Washiyama, N.; Bashar, A. H.; Terada, H.; Suzuki, K.; Yamashita, K.; Takinami, M. Role of biologic glue repair of proximal aortic dissection in the development of early and midterm redissection of the aortic root. *Ann. Thorac. Surg.* **2001**, *72*, 509–514.

16. García Páez, J. M.; Jorge-Herrero, E.; Carrera, A.; Millán, I.; Cordón, A.; Rocha, A.; Martín Maestro, M.; Castillo, J. L. Comparative study of the mechanical behaviour of a cyanoacrylate bioadhesive. *J. Mater. Sci. Mater. Med.* **2004**, *15*, 109–115.

17. Osmond, M. H. Pediatric wound management: the role of tissue adhesives. *Pediatr. Emerg. Care* **1999**, *15*, 137–140.

18. Osmond, M. H.; Klassen, T. P.; Quinn, J. V. Economic comparison of a tissue adhesive and suturing in the repair of pediatric facial lacerations. *J. Pediatr.* **1995**, *126*, 892–895.

19. Osmond, M. H.; Quinn, J. V.; Sutcliffe, T.; Jarmuske, M.; Klassen, T. P. A randomized, clinical trial comparing butylcyanoacrylate with octylcyanoacrylate in the management of selected pediatric facial lacerations. *Acad. Emerg. Med.* **1999**, *6*, 171–177.

20. Singer, A. J.; Thode, H. C., Jr. A review of the literature on octylcyanoacrylate tissue adhesive. *Am. J. Surg.* **2004**, *187*, 238–248.

21. Kaplan, M.; Bozkurt, S.; Kut, M. S.; Kullu, S.; Demirtas, M. M. Histopathological effects of ethyl 2-cyanoacrylate tissue adhesive following surgical application: an experimental study. *Eur. J. Cardiothorac. Surg.* **2004**, *25*, 167–172.

22. Timoshenko, S. In: Calpe, E. (Ed.) *Resistance of Materials* (Spanish version), Chapter I, pp. 1–6; Chapter X, pp. 274–285. Madrid, Spain, **1970**.

23. Sacks, M.; Chuong, C. J.; More, R. Collagen fiber architecture of bovine pericardium. *ASAIO J.* **1994**, *40*, M632–M637.

24. Garcia Páez, J. M.; Jorge, H. E.; Millán, I.; Rocha, A.; Maestro, M.; Castillo-Olivares, J. L.; Carrera, S. A.; Cordón, A. Resistance to tensile stress of a bioadhesive utilized for medical purposes: Loctite 4011. *J. Biomater. Appl.* **2004**, *18*, 179–192.

Chapter 10

SURFACE MODIFICATION OF CELLULOSE FIBERS WITH HYDROLASES AND KINASES

Tzanko Tzanov[1] and Artur Cavaco-Paulo[2]

[1]Departament d'Enginyeria Química, Universitat Politècnica de Catalunya, 08222 Terrassa, Spain
[2]Departamento de Engenharia Têxtil, Universidade do Minho, Campus de Azurém, 4800-058 Guimarães, Portugal

10.1 Introduction

Nowadays, the goods produced from cellulose fibers have the largest share in the textile market due to the excellent exploitation properties of these fibers. Cotton is breathable, moisture and heat conductor, drawing moisture and heat away from the body, thereby providing coolness at wearing and thus overall consumers' comfort. Cotton fibers are soft, hypoallergenic (do not irritate sensitive skin or cause allergies) and anti-static, which is why they are particularly favored for underwear and garments that get close to the skin, bandages and gauzes. Furthermore, the cellulose fibers can be easily blended with other fibers such as synthetics like polyester or lycra to combine the good mechanical properties of the synthetic polymers with the benefits of using natural-based materials in specific applications, e.g., medical textiles, easy-care fabrics, non-woven, composites, etc.

However, chemical modification of cellulose fibers by introducing new functional groups or compounds is still needed in order to improve their performance characteristics, e.g., dye fixation, water and soil-repellence, crease-resistance, handle, flame-retardance, and others. For this purpose, a large number of chemical compounds are used depending on the specific application of the fibers. This chapter refers to two distinct functional finishes for cellulose fibers, namely crease resistant and flame retardant.

J. V. Edwards et al. (eds.), Modified Fibers with Medical and Specialty Applications, 159–180.

10.1.1 Crease-resistance finishing

The crease-resistance of cellulose fibers is not satisfactory to meet the contemporary requirements for "easy-care" textile materials. When cotton fabric and garments are deformed, the numerous H-bonds, which exist between cellulose molecules are destroyed and consecutively created fixing the new deformed state of the fabric. The conventional crease-resistant finishes to improve the crease-resistance are mostly achieved on the basis of a condensation reaction at high temperature between *N*-methylol compound, such as dimethyloldihydroxyethylenurea (DMDHEU) or its derivatives and cellulose. In the last years, many efforts have been done to develop formaldehyde-free cross-linking agents since the formaldehyde, which is released during the fixation procedure has been identified as a potential human carcinogen. For this purpose, some product with lower formaldehyde content have been suggested—i.e., α-hydroxyalkylamides, polymers from urea and glutaraldehyde, and diamidodihydroxyethane products, all of them cross-linking the cellulose trough ether bond formation [1, 2]. Methylolamide adducts of formaldehyde and amides are common cross-linking agents for finishing of cellulosic fiber materials. The adducts of amides have large share in the pallet of conventional durable-press products for textile application [3]. One product of low formaldehyde content that became of commercial importance is the *N*-hydroxymethyl acryl amide. It reacts by *N*-methylol group with cellulose in a single step, using an acid-acting catalyst—zinc nitrate [4, 5]. The crease-resistance effect is provided by the formation of a network of ether bonds between each molecule of the reagent and cellulose (Scheme 10.1).

$$R_{cell}\text{-OH} + HO\text{-}CH_2\text{-}NH\text{-}CO\text{-}CH=CH_2 \longrightarrow R_{cell}\text{-}O\text{-}CH_2\text{-}NH\text{-}CO\text{-}CH_2\text{-}CH_2\text{-}O\text{-}R_{cell}$$

Scheme 10.1. Reaction between *N*-hydroxymethyl acryl amide and cellulose

Nowadays, only a few formaldehyde-free finishing agents are commercialized and are still quite expensive to replace entirely the traditional resins. The most promising formaldehyde-free reagents are the polycarboxylic acids—1,2,3,4-cyclopentanetetracarboxylic, 1,2,3,4-butanetetracarboxylic (BTCA), or combination of citric acid and polymers of maleic acid. The reaction with the cellulose substrate proceeds trough ester cross-linking, catalyzed by sodium hypophosphite [6, 7] (Scheme 10.2).

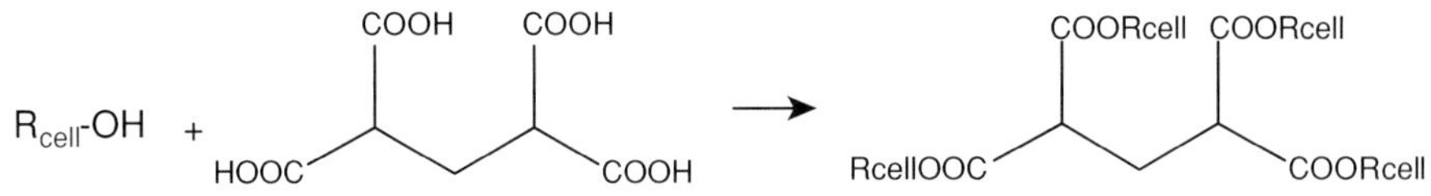

Scheme 10.2. Reaction between 1,2,3,4-butanetetracarboxylic acid and cellulose

It is a well-known drawback of any crease-resistance treatment on cotton fabrics that the improvement of the dimensional stability and wrinkle resistance is always correlated with significant decrease of the tensile strength [8, 9]. The loss of fabric mechanical strength has been attributed to two main factors: acid-catalyzed depolymerization and cross-linking of cellulose molecules [8, 10]. The cross-linking reduces the mobility of the macromolecules and greater stress, which could not be uniformly distributed along the polymer chains, is accumulated in fabric structure, leading to its disruption. The fabric strength loss caused by cross-linking is described as a reversible process and could be restored by removing chemically the cross-links [11–23]. However, the partial removal of the cross-links, so that the desired crease-resistance is preserved, requires the use of reagents that have minimal or controlled penetration beyond the fiber surface. Normally, the classical alkaline or acid hydrolysis (depending on the bonds to be broken) is difficult to be controlled and might damage seriously the crease-resistant effect and even the fabrics.

Currently, there is no commercially available enzymatic system, which specifically could attack the ether links between N-hydroxymethyl acrylamide and cellulose. However, several proteases have the potential to hydrolyze the amide bonds in N-hydroxymethyl acryl amide. The proteases catalyze the hydrolysis of certain peptide bonds in protein molecules. Serine-type proteases could act on the amide bonds in the cross-linked cellulose structure since these enzymes are known to tolerate large structural variations of the substrates and additionally are cheap and robust biocatalysts [24].

The ester bonds in polycarboxylic acids cross-linked cellulose could be partially hydrolyzed, restoring the strength of the fibers using lipases. The lipases are an unique class of hydrolases [25, 26], which are catalytically active in water, in mixtures of water and a water-immiscible organic solvent, and in organic solvent, catalyzing both hydrolysis and esterification reactions. The large excess of water favors the hydrolysis reaction. Most of the lipases are serine hydrolases, containing a serine residue in their active site, and forming intermediate acyl-enzyme substrate complex similarly to the above described proteases (Scheme 10.3).

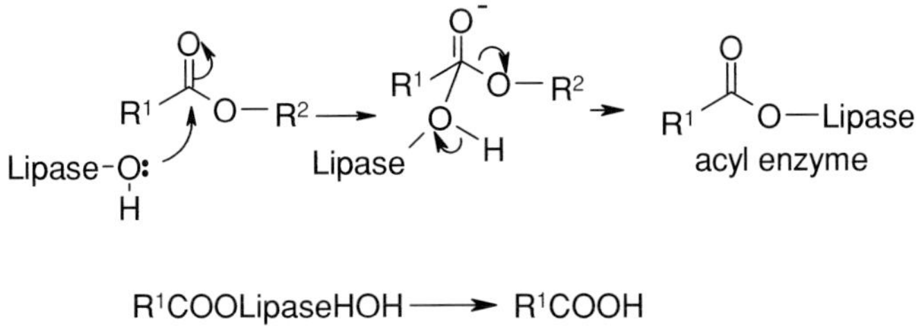

$$R^1COOLipaseHOH \longrightarrow R^1COOH$$

Scheme 10.3. Mechanism of the lipase catalyzed hydrolysis

10.1.2 Flame retardant finishing of cellulose fibers

Most of the currently commercialized products for flame retardant finishing of cellulose are based on phosphorous containing salts. However, the effect of flame retardance is not permanent since these products are not covalently fixed on the fibers and are gradually removed during washing. Conversely, chemical phosphorylation of cellulose usually is a rather complicated process, requiring several protection and deprotection steps [27, 28]. Enzymatic phosphorylation could make the synthesis more efficient, eliminating many of these steps. The introduction of phosphate groups in cellulose will provide permanent flame retardance and a highly reactive polymer toward various chemical types of compounds. The ability of hexokinase for phosphate group transfer suggested the application of this enzyme for functional modification of the C6-hydroxyl groups in cellulose (Scheme 10.4). Hexokinases catalyze phosphoryl transfer from adenosine-5′-triphosphate (ATP) [29–32] to the 6-hydroyl group of a number of furanose and pyranose compounds [33].

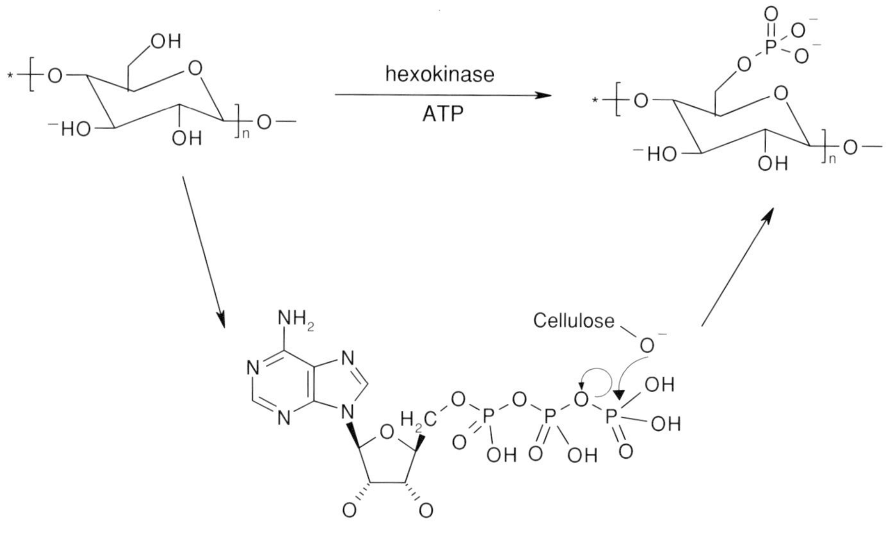

Scheme 10.4. Mechanism of hexokinase catalyzed cellulose phosphorylation

The objectives of this research were to perform a controlled biocatalytic hydrolysis of the amide bonds in *N*-hydroxymethyl acryl amide—and of the ester bonds in polycarboxylic acids—cross-linked cellulose fibers in order to restore partially the strength loss, without deteriorating the crease-resistance effect. On the other hand, enzymatic phosphorylation of cellulose would impart permanent flame-resistance to the fibers and will further provide functionalized cellulosic materials for broad range of applications.

10.2 Experimental

10.2.1 Textile material

The fabric used in this study was desized, scoured, and bleached cotton cellulose.

10.2.2 Crease-resistant finish

10.2.2.1 N-hydroxymethyl acryl amide application

The aqueous bath was prepared from 13% *N*-hydroxymethyl acryl amide (Aldrich) and 1.56% zinc nitrate hexahydrate (Aldrich) as catalyst. The cotton fabric was impregnated on foulard (80% wet pick-up), dried at 105 °C for 2 min and then cured in a Verner Mathis curing machine at 175 °C for 1.5 min.

10.2.2.2 1,2,3,4-Butanetetracarboxylic application

The fabric was impregnated on foulard (100% wet pick-up) with solution containing of 1,2,3,4-BTCA (from Aldrich), and anhydrous sodium hypophosphite (from Sigma) as the catalyst; with a BTCA-to-catalyst ratio of 3:2 (w/w) [15, 20]; dried at 80 °C for 5 min and then cured at 180 °C for 2 min. Afterward, the treated cellulose material was washed thoroughly in order to remove the remaining unreacted reagents and catalysts.

10.2.3 Enzyme treatment

10.2.3.1 Protease treatment

Serine protease (EC 3.4.21.62) from bacillus microorganism—Alcalase® 3.0 T (Novo Nordisk), 3 AU/g solid, was applied on *N*-hydroxymethyl acryl amide cross-linked cotton at pH 7.5 (0.1 M phosphate buffer) and 50 °C, in a shaker with 100 rpm [34].

10.2.3.2 Lipase treatment

Lipase (EC 3.1.1.3) Type VII from *Candida rugosa*, 724 U/mg solid, was purchased from Sigma and applied to the BTCA cross-linked fabrics at pH 7.5 (0.1 M phosphate buffer) and 37 °C, in a shaker with 100 rpm.

10.2.3.3 Hexokinase treatment

The enzymatic phosphorylation of the samples was carried out at 30 °C in an Ahiba Spectradye—Datacolor dyeing apparatus with closed vessels, at

40 rpm for 6 h, with 40 U/mL hexokinase (EC 2.7.1.1., Type IV: from Bakers Yeast, from Sigma; one unit will phosphorylate 1 μmol of D-glucose per min at pH 7.6 at 25 °C) dissolved in 50 mM phosphate buffer pH 7.6, and 50 mM deionized water solution of ATP, disodium salt (from Sigma), in liquor to fabric ratio 20:1. The fabrics were first treated for 15 min with the solution of ATP to ensure the impregnation of the textile material, and then the enzyme was added. After completing the process the enzymatically treated fabrics were washed at boil to remove any residual protein.

10.2.4 Analysis techniques

10.2.4.1 Wrinkle recovery angle and tensile strength

Wrinkle recovery angle (WRA) and tensile strength of the N-hydroxy methyl acryl amide and BTCA treated fabrics were measured according to AATCC test method 66-1990 and ASTM method D5035-90, respectively.

10.2.4.2 FT-IR analysis

Diffuse reflectance spectra of the samples were collected by Bomem MB-series FT-IR spectrometer, performing 100 scans for each spectrum. No smoothing function and baseline correction were applied. Potassium bromide was used to obtain the background spectrum. The effectiveness of the protease hydrolysis was assessed following the decrease of the amide carbonyl band intensity at 1667 cm^{-1} [35]. Lipase deesterification was assessed comparing the ester carbonyl band intensity at 1724 cm^{-1} [15–22]. The BTCA cross-linked fabrics were treated for 2 min at room temperature with 0.1 M NaOH to convert the free carboxyl to carboxylate to avoid overlapping of the carboxyl and the ester carbonyl bands [18].

10.2.4.3 Dyeability of cross-linked cellulose fibers

The protease and lipase treated samples (1 g each) were dyed 1 h, at 50 °C with 20 μM C.I. Acid Blue 25 and C.I. Basic Blue 9 (from Aldrich), respectively. The color of the fabrics was evaluated using a reflectance measuring Datacolor apparatus (LAV/Spec. Incl., d/8, D$_{65}$/10°) in terms of K/S values. The Kubelka–Munk relationship (K/S), where K is an adsorption coefficient and S is a scattering coefficient, is applied to textiles under the assumption that light scattering is due to the fibers, while adsorption of light is due to the colorant. The shift of the coordinates of the color in the cylindrical color space: L^*, a^*, and b^*, based on the theory that color is perceived by black–white (L), red–green (a), and yellow–blue (b) sensations, was summarized by the overall color difference (ΔE^*) value.

10.2.4.4 Detection of phosphorylation

Sigma procedure No. 345-UV for quantitative, ultraviolet, kinetic determination of glucose-6-phosphate dehydrogenase (G-6-PDH, EC 1.1.1.49) in
blood at 340 nm, was adopted for detection of glucose-6-phosphate (G-6-P)
formed in the hexokinase phosphorylation reaction. G-6-PDH is an enzyme,
which catalyzes the first step in the pentose phosphate shunt, oxidizing G-6-P
to 6-phosphogluconate (6-PG) and reducing nicotinamide adenine dinucleotide
phosphate (NADP) to NADPH, according to the reaction (Scheme 10.5):

$$\text{G-6-P} + \text{NADP}^+ \underset{\text{G-6-PDH}}{\overset{}{\longleftrightarrow}} \text{6-PG} + \text{NADPH} + \text{H}^+$$

Scheme 10.5. Reduction of NADP by G-6-PDH in the presence of G-6-P

NADP is reduced by G-6-PDH in the presence of G-6-P. The rate of formation of NADPH (measured spectrophotometrically following the increase in
absorbance at 340 nm) is proportional to the G-6-PDH activity and thus to the
concentration of G-6-P. The assay mixture contains 1 mL NADP (1.5 mmol/L)
and 0.01 mL G-6-PDH (2 U/mL; one unit will oxidize 1.0 μmol of D-glucose
6-phosphate to 6-phospho-D-gluconate per min in the presence of NADP at
pH 7.4 at 25 °C). The solution was stored at room temperature (18–26 °C) for
5–10 min. Then 2.0 mL G-6-P (1.05 mmol/L) was added and mixed gently.
The absorbance of the sample (A) at 340 nm versus water was recorded as A_i.
The final absorbance (A_f) was read and recorded 5 min later. The G-6-PDH
activity was defined as: $\Delta A \cdot \text{min}^{-1} = (A_i - A_f) \times 5^{-1}$. Standard curve of G-
6-PDH activity as function of G-6-P concentration was set ($Y = 0.292 \times X$,
$R^2 = 0.981$) so that the concentration of a sample, containing unknown amount
of G-6-P could be calculated there from. Then 1 g sample of phosphorylated
cotton fabric was completely hydrolyzed to the constituting glucose and presumably G-6-P units with 0.91 g/L total crude cellulase Ecostone L (from Röhm
Enzyme Finland), at 37 °C, pH 5, for 24 h. The solution was filtered trough
Ultrafree-4 centrifugal filter unit for concentration and purification of biological samples (from Millipore Corp.), to separate the protein from the sugars.
The sample of G-6-P in the Sigma assay was replaced with equivalent amount
of enzymatically hydrolyzed, hexokinase treated cellulose.

Dyeability of phosphorylated cellulose fibers

Samples of enzymatically phosphorylated cotton fabrics (0.5 g) were dyed
with 1.5 g/L of the following commercial grade dyes (Everlight Chem. Ind.
Corp.): Basic Green 4 at 60 °C, for 60 min; Reactive Blue 198, in the presence
of 60 g/L Na_2SO_4 and 20 g/L Na_2CO_3 (Sigma), at 80 °C, for 60 min; Direct
Red 224, in the presence of 0.1 g/L Na_2SO_4, at 90 °C, for 60 min. Each dyeing
experiment was repeated three times in an Ahiba Spectradye—Datacolor dyeing

Table 10.1. Calibration curves for the dyes used in the experiments

	Wavelength of maximum absorbance	Equation	R^2
C.I. Direct Red 224	521 nm	$Y = 19.01 \times X$	0.9987
C.I. Basic Green 4	616 nm	$Y = 142.27 \times X$	0.9918
C.I. Reactive Blue 198	629 nm	$Y = 18.51 \times X$	0.9928

apparatus with liquor to textile material ratio 100:1. The hexokinase-treated fabrics were dyed also with 13.4 g/L Disperse Red 60, in the presence of 30 g/L H_2NCN (50%) from Aldrich. The fabrics were padded on foulard (80% wet pick-up), pre-dried at 110 °C for 1 min, and then thermofixed with dry air in a Werner-Mathis curing machine, at 190 °C for 3 min. Dyed fabrics were repeatedly washed-off at boil to remove the unfixed dye. Improvement of the dyeing results would be an indirect experimental confirmation for the occurrence of the phosphorylation reaction.

Rate of dyes exhaustion and fixation. In order to define the rate of exhaustion, samples were taken from the dyeing liquors every 10 min during the dyeing, and were studied spectrophotometrically (Unicam Heλios UV-Vis spectrophotometer) after adequate dilution at the corresponding wavelength of maximum absorbance for the different colorants. The concentrations of the baths were calculated according to previously set dye calibration curves (Table 10.1).

The percentage of exhaustion was determined as a proportion between the concentration of the dye initially presented in the dyebath—C_i, and the dye remaining at the end of dyeing—C_f (Eq. 1):

$$E(\%) = \left(1 - \frac{C_f}{C_i}\right) \times 100 \qquad (1)$$

10.3 Results and discussion

10.3.1 Proteases to improve the strength of *N*-hydroxymethyl acryl amide cross-linked cellulose

10.3.1.1 Duration of the enzymatic hydrolysis

The time-dependent effect of the protease product (0.5 g/L) on the properties of the cross-linked fabric is depicted in Figure 10.1. The strength loss recovery of the fabrics, due to the protease treatment, reached about 15% after 30 min of enzyme reaction. Thus, nearly one half of the strength loss caused by the

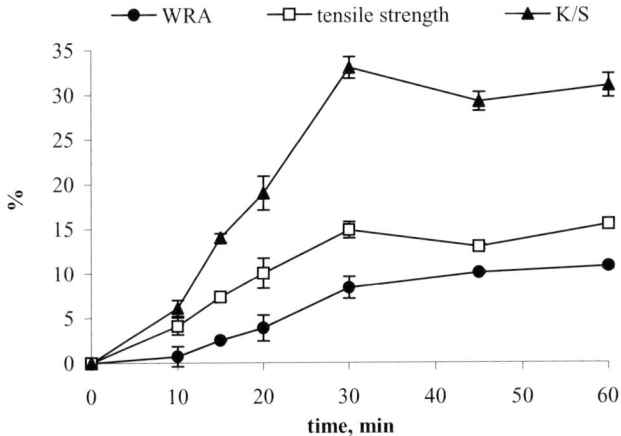

Figure 10.1. Time dependence of tensile strength recovery, WRA decrease and acid dye dyeability of crease-resistant finished cotton (13% *N*-hydroxymethyl acryl amide/1.5% Zn(NO$_3$)$_2$.6H$_2$O) hydrolyzed by 0.5 g/L protease

cross-linking was recovered during the enzymatic hydrolysis. Prolonging the reaction time did not provide further strength improvement. Concomitantly the WRA decreased by 8%. These changes in the mechanical characteristics of the fabrics occurred during the first 30 min of reaction. The short time, sufficient to improve the tensile strength of the fabrics, renders the enzymatic process suitable for continuous operations in textile practice.

Comparative alkaline hydrolysis of the cross-linked cotton, carried out with 0.1 M NaOH, at 50 °C for 30 min did not provide better recovery of the mechanical strength (about 10%), however, caused twice higher loss of the crease-resistant effect (about 15%) compared to the enzymatic process (Table 10.2).

The cleavage of the amide bonds in the cross-linked cotton was expected to produce carboxylic and amine end groups. Thus, the dyeability with an acid

Table 10.2. Tensile strength and WRA of cotton fabric treated with 13% *N*-hydroxymethyl acryl amide/1.5% Zn(NO$_3$)$_2$.6H$_2$O and hydrolyzed by NaOH or protease for 30 min

Fabric	Wrinkle recovery angle, [°], ($w + f$)	Strength, [N]
Blank	129	590
Cross-linked	291	405
Cross-linked and enzymatically treated	266	465
Cross-linked and alkaline treated	242	445

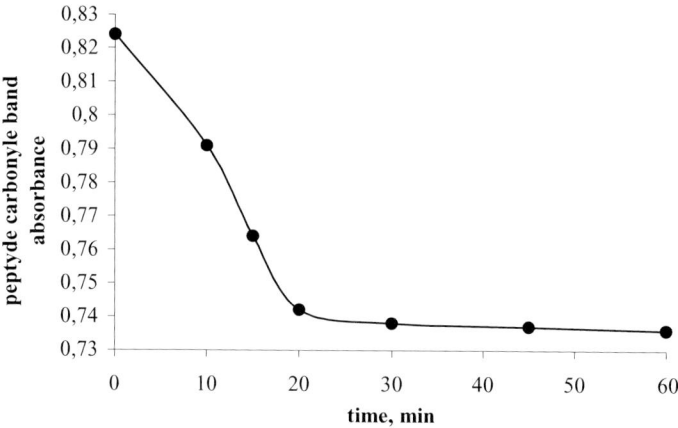

Figure 10.2. Time dependence of amide carbonyl band absorbance at $1667\,cm^{-1}$ for cotton, cross-linked with 13% *N*-hydroxy methyl acryl amide/1.5% $Zn(NO_3)_2.6H_2O$ and then hydrolyzed by 0.5 g/L protease

dye (C.I. Acid Blue 25) could be considered as an indirect indication for the occurrence of hydrolysis. The K/S values (dyeability) of enzymatically treated fabric notably increased (by 30%) compared to the K/S values for the non-hydrolyzed one (Figure 10.1). The increased affinity of the protease-hydrolyzed fabric toward an acid dye may be interpreted as due to the appearance of free amine groups, coming from the amide bonds cleavage. After 30 min, the dyeability of the enzymatically treated fabrics reached saturation.

The intensity of the amide carbonyl peak at $1667\ cm^{-1}$ decreased with the increase of the hydrolysis time (Figure 10.2). Most significant decrease of the signal was observed for reaction times up to 20 min. The carbonyl band intensity reached a plateau after 30 min of enzymatic treatment, e.g., no further hydrolysis occurred.

When considering enzymes that could hydrolyze peptides it should be noticed that the amide bond is very stable (e.g., in contrast to esters, particularly when exposed to water), presenting a planar, resistant to deformation structure. The amide hydrolysis proceeds through two steps—acylation and deacylation, where the formation of acyl-enzyme intermediate is the rate-limiting step (Scheme 10.6). A property, shared by all serine proteases is the hydrogen-bonded triad of Asp-His-Ser. The enzyme is acetylated on the Ser residue. An important feature in the catalytic reaction is the formation of an ester between the oxygen of serine and the "acyl" portion of the substrate, with release of the "amino" portion of the substrate as the first product, P_1. For the release of the second product P_2 reaction with water is necessary. Since, more than one amide group in the crease-resistant finished cotton fabric might react with the enzyme, many end amino and carboxylic groups may appear. When a nucleophile, such

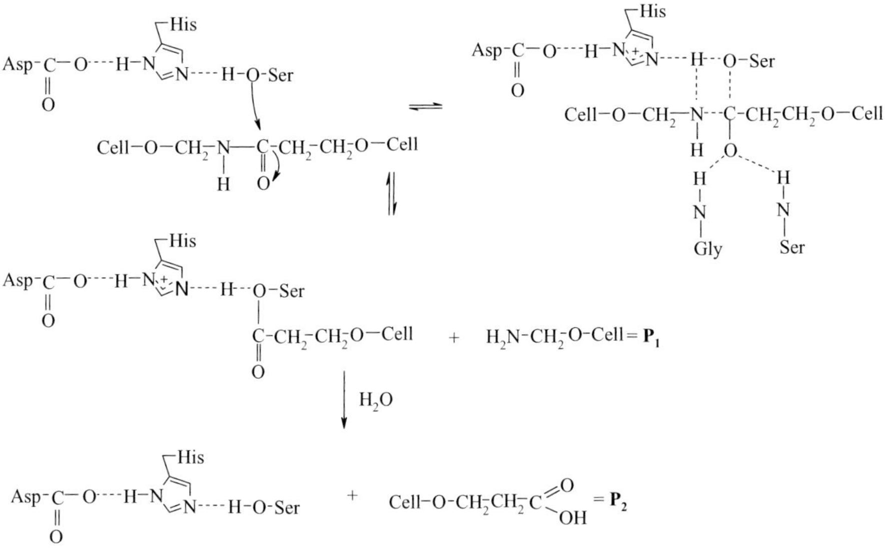

Scheme 10.6. Protease hydrolysis of amide cross-linked cotton

as amine, is present in the reaction between protease and substrate, a transacylation reaction with the amine might occur and the rate of the disappearance of the substrate becomes steady, e.g., further hydrolysis would be inhibited. The amide bonds in the cellulose fibers cross-linked with N-hydroxy methyl acryl amide are not a typical substrate for the enzyme. In contrast to low molecular substrates, the products of the enzyme hydrolysis remain on the fabric since the cross-linking agent is still bound to the cellulose *via* stable ester bonds, which are not susceptible to hydrolysis. Perusal of the data presented in the above Figures 10.1 and 10.2, and Table 10.2 supports the hypothesis that with the proceeding of the enzyme hydrolysis of the cross-linked amide product, the number of the nucleophilic amine groups should increase and thus the transacylation reaction would become predominant. Saturation was observed in the tensile strength recovery and in the dyeability of the fabrics. Another explanation for the saturation of the enzymatic effect might be the limited accessibility of the cross-links to the large enzyme molecules. The WRA did not change proportionally to the improvement in strength, due to the fact that the linker still remained on the fibers.

10.3.1.2 Protease concentration

The effect of the enzyme concentration on the strength loss recovery and WRA was studied for 30 min of reaction. As it can be seen in Figure 10.3, the tensile strength recovery normally increases with the increase of the concentration.

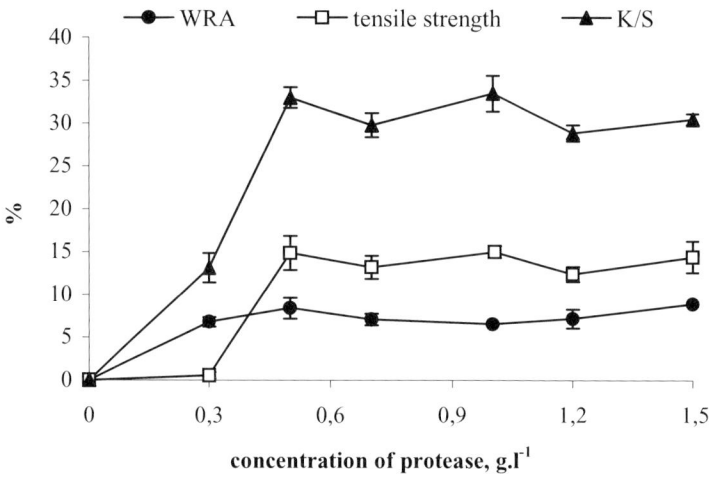

Figure 10.3. Effect of protease concentration on the tensile strength recovery, WRA, and acid dye dyeability at 30 min reaction time

At 0.5 g/L protease, the strength recovery reached approximately 15%. Above this concentration the amount of enzyme did not cause any further improvement of fabric tensile behavior. The decrease of the WRA was up to 9% and followed the same concentration dependence. The dyeability of the fabrics with an acid dye increased rapidly for protease concentrations up to 0.5 g/L and then remained steady independently on the amount of enzyme. The inhibition of the enzyme hydrolysis is reflected as well in the decrease of the intensity of the amide carbonyl peak at 1667 cm^{-1} in the infrared spectrum of the fabrics (Figure 10.4).

10.3.2 Lipases to improve the strength of BTCA cross-linked cellulose

10.3.2.1 Lipase concentration

The effect of the enzyme concentration on tensile strength and WRA of BTCA treated fabrics was investigated. Tensile strength normally increases as the hydrolysis of the cross-linked fabric progresses. In our experiment, the tensile strength of the BTCA cross-linked fabrics increased when up to 0.4 g/L lipase was applied for hydrolysis of the ester bonds. Above this enzyme concentration further improvement of fabrics tensile performance was not observed (Figure 10.5). This limitation of the hydrolysis could be explained with the restricted accessibility to the enzymatic attack of the BTCA/cellulose ester crosslinks in the dense cellulose structure. The ester groups have a slightly higher concentration in the interior of the fabric than in its near surface [16]. The large enzyme molecule is not able to enter in contact with substrate and to form the

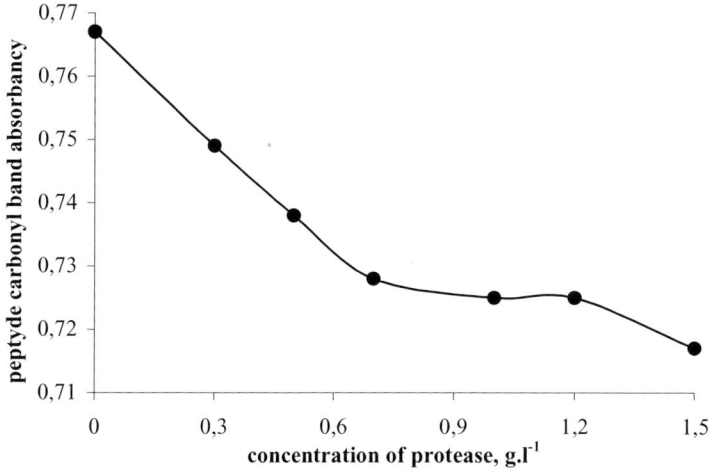

Figure 10.4. Amide carbonyl band absorbance at 1667 cm^{-1} of cotton cross-linked with 13% *N*-hydroxymethyl acryl amide/1.5% Zn(NO$_3$)$_2$.6H$_2$O and then hydrolyzed by different concentration of protease for 30 min

intermediate enzyme–substrate complex. The size of the protein molecule is a self-limiting factor for undesirable further hydrolysis of the BTCA/cellulose cross-linkages and lessening of the crease-resistance effect.

It is known that the electrostatic enzyme–substrate interactions play an important role in the enzyme catalysis. It was found that the active site of *Candida rugosa* lipases becomes negatively charged at pH 6 [36] (in our

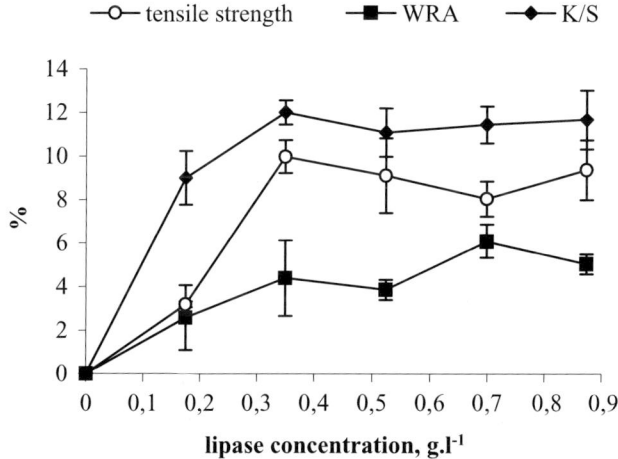

Figure 10.5. Tensile strength recovery, WRA decrease and decrease in dyeability with cationic dye of BTCA treated cotton (6% BTCA/4% NaH$_2$PO$_2$), hydrolyzed by different concentration of lipase for 1 h

experiments the lipases were applied at pH 7.5). The negative charge in the active site is needed for the expulsion of the negatively charged hydrolysis products and is therefore essential for the maximum activity of the enzyme. After cleavage of the ester bond, an ionized acid will be ejected from the active site due to electrostatic repulsion. If the active site is positively charged, a product inhibition would occur due to interaction of the active site with the negatively charged polycarboxylic acid. In the case of cross-linking of cellulose with polycarboxylic acids, four ester bonds per acid molecule are possible. Even though the enzyme hydrolyzed some of these ester cross-linkages, the product still remains on the fabric due to singly bonded acid side groups [16]. Thus, hydrolysis of the ester groups is faster than the rate of BTCA molecule removal. With the progress of the hydrolysis reaction the fabric will become more negatively charged and the enzyme will be rejected from fabric's surface, impeding further hydrolysis. As more hydrolysis occurs, the intermediate binding between enzyme and substrate becomes more difficult due to electrostatic repulsion. This might be another explanation, apart from the steric constraints, why did the hydrolysis reach a maximum and afterward remained steady independently on the enzyme concentration.

A comparative mild chemical hydrolysis was performed using 0.1 M NaOH, at 37 °C for 1 h. The alkaline hydrolysis was accompanied with considerable loss of the desired crease-resistant effect—nearly 40% decrease of WRA versus 30% of strength improvement. At the same conditions, samples treated with enzyme showed about 4% decrease of WRA, which corresponds to nearly 10% strength recovery (Figure 10.5). Even though BTCA molecules were gradually removed from the fabric as a result of hydrolysis, the effectiveness of the remaining BTCA molecules as a cross-linking agent decreased only slightly. The change of WRA was twice lower than the tensile strength recovery of BTCA cross-linked cotton.

The dyeability of BTCA treated fabrics with Basic Blue 9 decreased after 1 h hydrolysis with up to 0.4 g/L lipase (Figure 10.5). Significant correlation could be found between the tensile strength recovery and the decrease in dyeability of the hydrolyzed cotton. We expected that the presence of free carboxyl and carboxylate groups on the BTCA treated cotton after the cleavage of the ester bonds would make it dyeable with basic dyes. However, the removal of the product during the enzymatic hydrolysis reduces the affinity of the cationic dye toward the fabric. The alkali hydrolyzed BTCA cotton was worst dyed (about 50% decrease of K/S), while the untreated cotton was not dyeable with cationic dye.

The infrared spectroscopy data in Figure 10.6 showed decrease of the intensity of the ester carbonyl peak at 1724 cm^{-1} with increase of the lipase concentration. Above 0.4 g/L lipase, where highest tensile strength recovery was observed, the curve of the carbonyl band intensity reached a plateau, e.g., no further hydrolysis occurred.

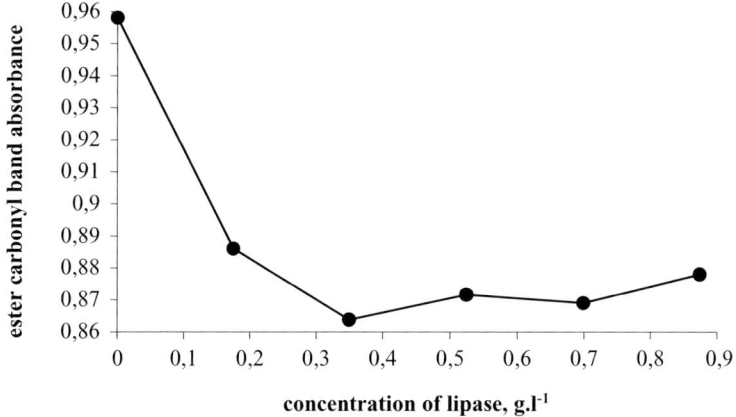

Figure 10.6. Ester carbonyl band absorbance at 1724 cm^{-1} of cotton treated with 6% BTCA/4% NaH$_2$PO$_2$ and hydrolyzed with different concentrations of lipase

10.3.2.2 Hydrolysis time

The effect of the hydrolysis duration on fabric tensile performance and crease-resistance was examined at fixed lipase concentration (0.4 g/L). The time of process was varied from 30 min to 25 h. The tensile strength recovery reached a maximum during the first hour of the process and remained constant for the next 24 h of treatment (Figure 10.7). The decrease of WRA followed the same tendency. Inactivation of the enzyme by the hydrolysis products and long operational time is equally possible, apart from the above stated steric

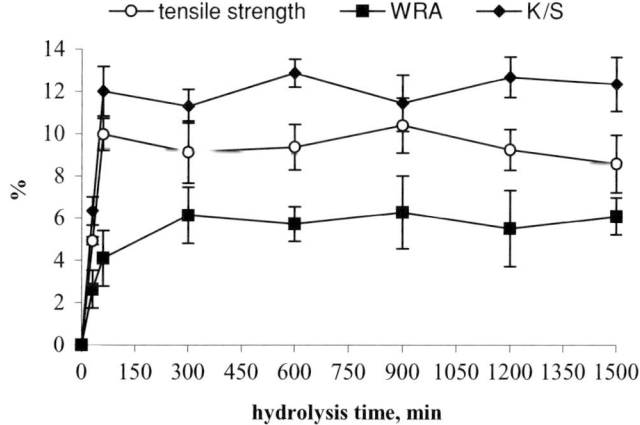

Figure 10.7. Tensile strength recovery, WRA decrease and decrease in dyeability with cationic dye of BTCA treated fabrics (6% BTCA/4% NaH$_2$PO$_2$), hydrolyzed by 0.4 g/L lipase for different times

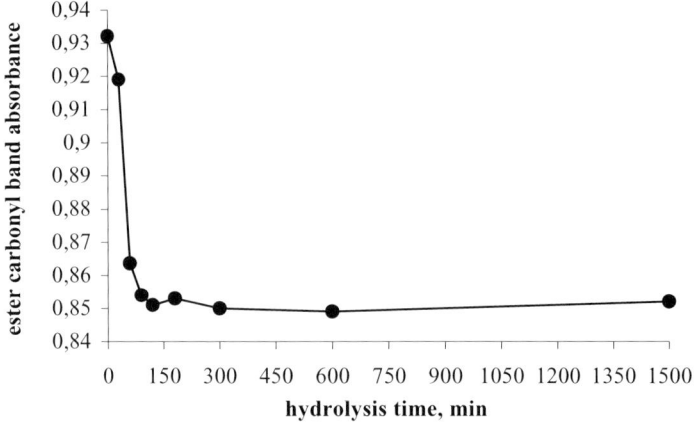

Figure 10.8. Ester carbonyl band absorbance at 1724 cm^{-1} of cotton treated with 6% BTCA/4% NaH$_2$PO$_2$ and hydrolyzed with 0.4 g/L lipase for different times

and electrostatic considerations. The decrease in intensity of the ester carbonyl peak in the infrared spectrum of the fabric corresponds to inhibition of the enzymatic hydrolysis (Figure 10.8).

10.3.3 Phosphorylation of cotton cellulose with baker yeast hexokinase

The spectrophotometrically detected G-6-PDH activity with the sample of hydrolyzed hexokinase treated fabric is a clear evidence for the occurrence of enzymatic phosphorylation of cellulose. This enzymatic approach provides an alternative to the ^{31}P NMR or XPS techniques, since the degree of phosphorylation expected on the fabrics is rather low. The enzymatic procedure is reliable due to the high specificity of the enzyme toward the substrate under investigation. The concentration of the G-6-P (0.01986 g/L) in the cellulose hydrolisate was calculated according to the previously set calibration curve. This value corresponds to phosphorylation of 0.03% of the glucopyranose units. The phosphorylation occurs on the primary alcohol groups of cellulose, which are most available for reaction. The secondary 3 OH group is involved with the ring oxygen atom in extra hydrogen bonds, and the chain is additionally coiled, protecting the 2 OH. Conventional dyeing procedures [37] with basic, direct, and reactive dyes resulted in up to 10% improvement in dye fixation on hexokinase treated cotton (Figures 10.9–10.11). Enzymatically phosphorylated fabrics retarded nearly threefold the fire propagation [38] (1.1 cm^2/sec), compared to the untreated cotton samples (3 cm^2/sec), when exposed to direct flame (Figure 10.12). These results are illustrated by the examples below.

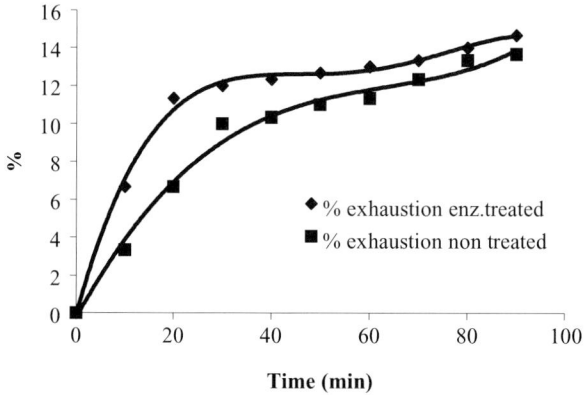

Figure 10.9. Rate of dyebath exhaustion of Basic Green 4, determined according to its calibration curve set at the wavelength of maximum dye absorbance, 616 nm

10.3.3.1 Dyeing with basic (cationic) dyes

Cationic dyes are primarily applied to acrylics, binding to the anionic acrylic groups at higher than the glass transition temperature of the polymeric substrate. Though the cationic dyes are not originally for dyeing of cotton, the higher dye up-take on phosphorylated cellulose is a clear indication for the increased negative charge of the textile substrate, due to the presence of phosphate groups (Figure 10.9).

10.3.3.2 Dyeing with direct (anionic) dye

In the case of dyeing of cellulose fibers with direct dyes the exhaustion is a measure of the substantivity of the anionic dye toward the fiber substrate, provided sufficient time is allowed for the sorption equilibrium to be reached. The fixation of the direct dyes is governed by their lengthy planar structure trough H-bonds and dipole interactions with the cellulose polymer. The fixation yield on hexokinase treated fabrics was higher than on the untreated ones (Figure 10.10) as a result of the increased polarity of the cellulose substrate.

10.3.3.3 Dyeing with reactive dye

The large liquor to goods ratio (100:1) used in the experiment favored the hydrolysis of the dye and thereby reduced the dyeing efficiency, which could explain the relatively low dye exhaustion and fixation level. Nevertheless, higher dye fixation was achieved on the hexokinase-phosphorylated cellulose due to the introduction of more active centers (more OH groups) for ester bond formation with the monochlorotriazine reactive dye (Figure 10.11).

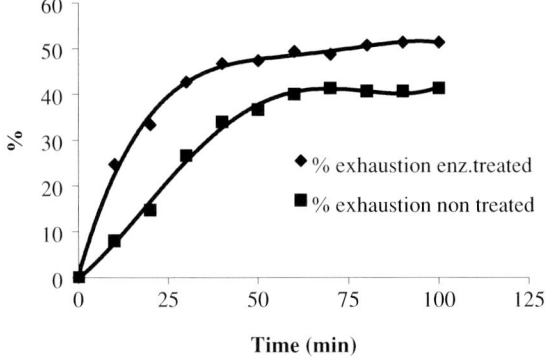

Figure 10.10. Rate of dyebath exhaustion of Direct Red 224, determined according to its calibration curve set at the wavelength of maximum dye absorbance, 521 nm

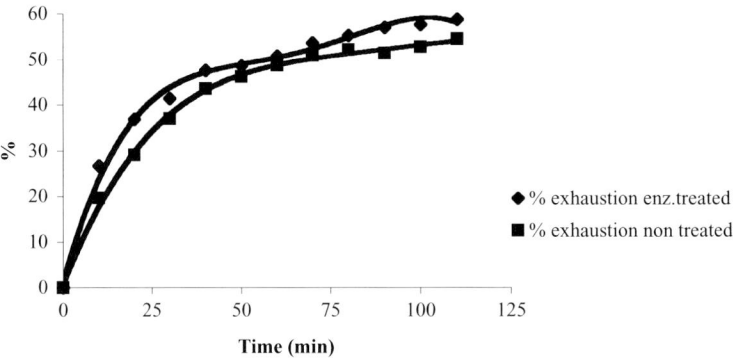

Figure 10.11. Rate of dyebath exhaustion of Reactive Blue 198, determined according to its calibration curve set at the wavelength of maximum dye absorbance, 629 nm

10.3.3.4 Dyeing of cotton fabrics with disperse dyes

Most of the reactive dyes on cellulose are fixed under alkaline conditions. However, there is another class of reactive dyes, which are applied, from mildly acidic solution—the phosphonic acid dyes [39–41]. The synthesis of these dyes originated from earlier research on durable flameproofing of cotton by treatment with phosphonic acid, or methan phosphonic acid and cyanamide-type compounds. Bond formation with the substrate was achieved at high temperature 200–220 °C, for 1–3 min, through pad-dry-thermofix dyeing procedure. The exact mechanism of the dye–fiber interaction is so far not fully understood, and at least two possible pathways for dye fixation were suggested. In general, the following simplified scheme (Scheme 10.7) of the reaction is assumed:

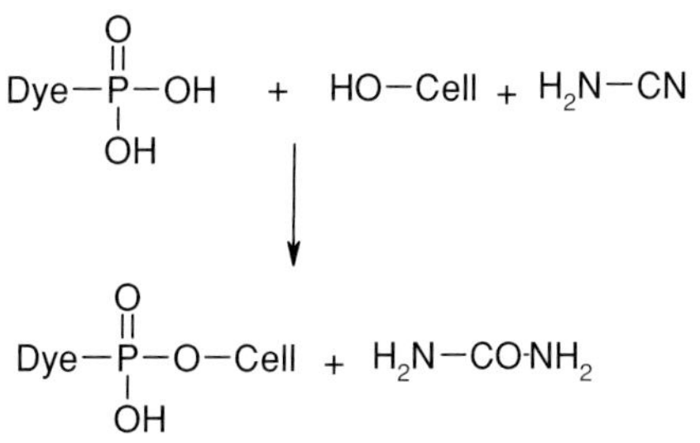

Dye—P(=O)(OH)—OH + HO—Cell + H₂N—CN

$$\text{Dye}-\overset{\overset{\displaystyle O}{\|}}{\underset{\underset{\displaystyle OH}{|}}{P}}-OH \quad + \quad HO-Cell \quad + \quad H_2N-CN$$

$$\downarrow$$

$$\text{Dye}-\overset{\overset{\displaystyle O}{\|}}{\underset{\underset{\displaystyle OH}{|}}{P}}-O-Cell \quad + \quad H_2N-CO\cdot NH_2$$

Scheme 10.7. Generalized mechanism of the reaction of cotton fabrics with phosphonic dyes in the presence of cyanamide

The new type of phosphorylated cellulose, bearing already the reactive group, otherwise coming from the dye, suggested that the modified cellulose could react with disperse dyes, which do not possess reactive groups, but just OH groups, in the presence of cyanamide, in a similar way as in the case of the acid phosphonic dyes (Scheme 10.8):

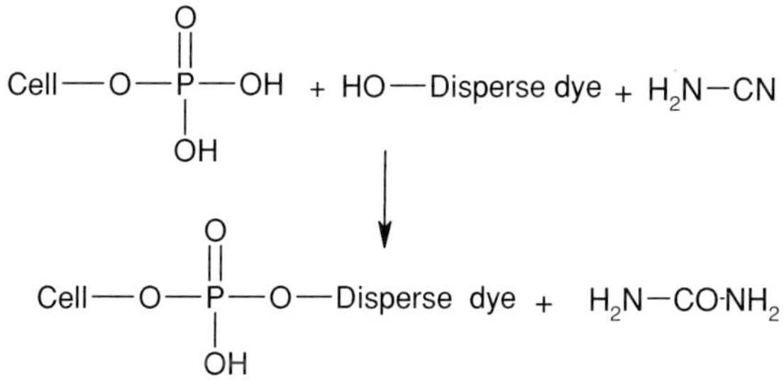

$$\text{Cell}-O-\overset{\overset{\displaystyle O}{\|}}{\underset{\underset{\displaystyle OH}{|}}{P}}-OH \quad + \quad HO-\text{Disperse dye} \quad + \quad H_2N-CN$$

$$\downarrow$$

$$\text{Cell}-O-\overset{\overset{\displaystyle O}{\|}}{\underset{\underset{\displaystyle OH}{|}}{P}}-O-\text{Disperse dye} \quad + \quad H_2N-CO\cdot NH_2$$

Scheme 10.8. Expected mechanism of the reaction of hexokinase-treated cellulose fibers with disperse dye in the presence of cyanamide

The deeper color shade obtained on hexokinase treated fabrics, dyed with disperse dyes (Table 10.3) is a confirmation of the proposed dye-cellulose reaction mechanism, with the participation of phosphate groups, belonging to the cellulose. The untreated fabric was poorly dyed, because the disperse dyes are specifically designed for polyester fibers and possess low affinity toward cellulose; except for if reactive cites on the fibers were created.

Table 10.3. Color difference between the untreated and the enzymatically treated samples dyed with Disperse Red 60, in the presence of cyanamide

ΔE^*	ΔL^*	Δa^*	Δb^*
7.368	4.484	−5.834	0.380

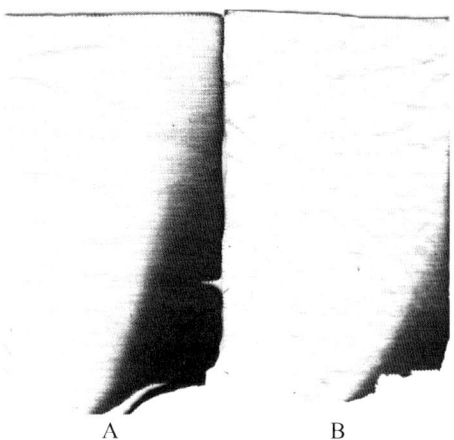

Figure 10.12. Illustration of the results from the test for flame-resistance: (A) untreated
and (B) hexokinase treated fabrics

10.3.3.5 Flame-resistance of hexokinase treated cotton fabrics

Durable flame-resistance of textile fiber materials is achieved traditionally
by cross-linking with reactive compounds containing phosphonium structures.
The decrease in flammability of the hexokinase-phosphorylated cellulose supports the occurrence of the enzymatic functional modification (Figure 10.12).

10.4 Conclusions

Hydrolases and kinases based enzymatic processes, performed at mild to
the cellulose fibers conditions—neutral pH and moderate temperature were
efficient to restore the tensile strength of resins cross-linked cellulose fibers
and to introduce new phosphate functional groups in fiber structure.

The protease hydrolysis of the amide bond in N-hydroxymethyl acryl amide
cross-linked cellulose resulted in about 15% strength loss recovery coupled
with up to 8% decrease of the crease-resistance effect. The lipase treatment of
BTCA cross-linked cotton fabrics improved the tensile strength with up to 10%,
causing just 4% deterioration of the wrinkle-resistance. Data from FT-IR spectroscopy confirmed the occurrence of protease and lipase catalyzed hydrolysis
registering a decrease of the intensity of the amide and carboxyl carbonyl peaks
in enzymatically treated samples. The size of the enzymes and related steric
difficulties to form the enzyme–substrate intermediate complex restricted the
process on fiber surface preventing further undesirable cross-links removal.
In contrast, in the conventional alkaline hydrolysis the deterioration of the
wrinkle-resistance of the fabrics predominates on the recovery of the strength
loss.

Cotton cellulose was enzymatically phosphorylated in a new biosynthetic process using hexokinase in the presence of phosphoryl donor ATP. An innovative enzymatic analytical approach, originally developed for determination of G-6-PDH, was used to detect the occurrence of phosphate functional modification of cellulose. Phosphorylation of 0.03% of the glucopyranose units in the cellulose fibers provided a textile material with improved dyeability and flame-resistance, which additionally might be used for large number of chemical synthesis.

References

1. Frick, J.; Harper, R. *Textile Res. Inst.* **1982**, *2*, 141.
2. Lämmermann, D. *Melliand Textilberichte*, **1992**, *73*, 274.
3. Frick, J.; Harper, R. *Textile Res. Inst.* **1982**, *2*, 141.
4. Manly, R. H. *Crease resistant Treatment of Fabrics*, Park Ridge, New Jersey, **1976**, 104.
5. U.S. Patent 3,561,916, **1971**, United States of America, W. F. Britinger.
6. U.S. Patent 4,820,307, **1989**, United States of America, Department of Agriculture, invs.: C. Welch, B. Andrews.
7. Welch, C. *Textile Res. J.* **1988**, *8*, 480.
8. Meyer, U.; Mueller, K.; Zollinger, H. *Textile Res. J.* **1976**, *46*, 813.
9. Murphy, A. L.; Margavio, M. F.; Welch, C. M. *Textile Res. J.* **1971**, *41*, 22.
10. Zeronian, S. H.; Bertoniere, N. R.; Alger, K. W.; Duffin, K. W.; Kim, M. S.; Dubuque, L. K.; Collins, M. J.; Xie, C. *Textile Res. J.* **1989**, *59*, 484.
11. Kang, I.-S.; Yang, C. Q.; Wei, W.; Lickfield, G. *Textile Res. J.* **1998**, *68*, 865.
12. Kang, I.-S.; Yang, C. Q.; Wei, W.; Lickfield, G. *Textile Res. J.* **1998**, *68*, 865.
13. Wei, W.; Yang, C. Q.; Jiang, Y. *Textile Chem. Color.* **1999**, *31*, 34.
14. Wei, W.; Yang, C. Q. *Textile Res. J.* **1999**, *69*, 145.
15. Yang, C. Q.; Bakshi, G. *Textile Res. J.* **1996**, *66*, 377.
16. Yang, C. Q. *Textile Res. J.* **1991**, *61*, 298.
17. Yang, C. Q. *Textile Res. J.* **2001**, *71*, 201.
18. Yang, C. Q. *Textile Res. J.* **1991**, *61*, 433.
19. Yang, C. Q.; Mao, Z.; Lickfield, G. *Text. Chem. Color.* **2000**, *32*, 43.
20. Yang, C. Q.; Wang, X. *Textile Res. J.* **1996**, *66*, 595.
21. Yang, C. Q.; Wang, X.; Kang, I.-S. *Textile Res. J.* **1997**, *67*, 334.
22. Yang, C. Q.; Xu, L.; Li, S.; Jiang, Y. *Textile Res. J.* **1998**, *68*, 457.
23. Yang, C. Q.; Wang, D. *Textile Res. J.* **2000**, *70*, 615.
24. Cavaco-Paulo, A. *Enzyme Applications For Fiber Processing*, ACS Symposium Series Book, **1998**, *687*, 36.
25. Alberghina, L.; Schmid, R. D.; Verger, R. (Eds.), *Lipases: Structure, Mechanism and Genetic Engineering*. Weinheim, VCH, **1990**.
26. Borgström, B.; Brockman, H. L. *Lipases*. Elsevier, Amsterdam, **1984**.
27. Edwards, J. V.; Yager, D. R.; Cohen, I. K.; Diegelmann, R. F.; Montante, S.; Bertoniere, N.; Bopp, A. F. *Wound Repair Regen.* **2001**, *1*, 50.
28. Kim, S. S.; Jeong, W. Y.; Shin, B. C.; Oh, S. Y.; Rhee, J. M. *J. Biomed. Mater. Res.* **1998**, *40*, 401.
29. Chenault, H. K.; Simon, E. S.; Whitesides, G. M. *Biotechnol. Genet. Eng. Rev.* **1988**, *6*, 221.
30. Whitesides, G. M.; Wong, C. H. *Aldrichimica Acta*, **1983**, *16*, 27.

31. Whitesides, G. M.; Wong, C. H.; Pollak, A. *ACS Symp. Ser.* **1982**, *185*, 205.

32. Crans, C.; Kazlauskas, R. J.; Hirschbein, B. L.; Wong, C. H.; Abril, O.; Whitesides, G. M. In: Mosbach, K. (Ed.) *Methods Enzymol.* Academic Press Inc., New York, **1987**, *136*, 263.

33. Viola, R. E.; Raushel, F. M.; Rendina, A. L.; Cleland, W. W. *Biochemistry*, **1982**, *21*, 1295.

34. "*Alcalase*® *manual instruction*", Enzyme Business, B259f-GB Nov. **1998** © Novo Nordisk A/S.

35. Silverstein, R. M.; Bassler, G. C.; Morril, T. C. *Spectrometric Identification of Organic Compounds*, 5th edn. John Wiley & Sons Inc., Toronto, **1991**, 102–130.

36. Petersen, M.; Fojan, P.; Petersen, S. *J. Biotechnol.* **2001**, *85*, 115.

37. Giles, C. H. *A Laboratory Course in Dyeing*, The Society of Dyers and Colourists, Bragford, Yorkshire, England, **1974**, 65.

38. Norme française NF G 07-184, Textiles-Comportement au feu, Méthode de classement en fonction de la surface br"ûlée, **1985**.

39. Renfrew, A. H. M.; Taylor J. A. *Rev. Prog. Coloration*, **1990**, *20*, 1–9.

40. O Nkeonye, P. *JSDC*, **1986**, *102*, 384–391.

41. Amato, M. E.; Fisichella, S.; Rattee, I. D. *JSDC*, **1987**, *103*, 434–437.

Chapter 11

ENZYMES FOR POLYMER SURFACE MODIFICATION

G. Fischer-Colbrie[1], S. Heumann[2], and G. Guebitz[1,2]

[1]*Applied Biocatalysis Research Center, Steyrergasse 17/6, A-8010 Graz, Austria*
[2]*Department of Environmental Biotechnology, Graz University of Technology, Petersgasse 12, A-8010 Graz, Austria*

Abstract

During the recent years, various bacteria and fungi have been investigated concerning their ability to degrade artificial polymers. This chapter reviews recent work on enzymes used for hydrolysis/oxidation of synthetic polymers and discusses their use in polymer surface modification.

11.1 Introduction

Polyethylene terephthalate (PET), polyamide (PA), and polyacrylonitrile (PAN) are the most widely used polymers in textile industry. Manmade fibers in general have a market share of 54.4% in textile industry and tend to gain even more share compared to natural fibers. The advantages of manmade fibers over natural fibers range from a lower specific weight to faster drying, better water transport properties and to better easy-to-clean-properties.

The following Table 11.1 comprises the worldwide production of polymers used for textile manufacturing.

Despite the increase in the production volume, the market shows a desire for improved synthetic textile properties like wettability or increased hydrophilicity. Furthermore, effects like better dyeability with water-soluble dyes or surface functionalization for special applications like coupling of flame retardants are desired. Strategies to obtain improvement in the said parameters range from chemical treatment (e.g., alkali treatment) to plasma treatment (e.g., low-temperature plasma treatment) [2, 3]. When compared to chemical treatments,

J. V. Edwards et al. (eds.), Modified Fibers with Medical and Specialty Applications, 181–189.
© 2006 *Springer. Printed in the Netherlands.*

Table 11.1. Worldwide production of polymers used in textile industry [1]

Polymer	Yearly production (2001)	±% compared to 2000
Polyester	31.3	+3.1
Polyamide	3.7	−8.7
Polypropylene	3.2	2.9
Polyacrylonitrile	2.6	−2.0

an enzymatic treatment of the surface would have the advantage of a reaction limited to the polymer surface only, and milder treatment conditions leading to less damage of the fibers. It may save energy and avoids the use of strong chemicals, which can become a problem when discharged with the process effluent. Oxidative enzymes or mediator-triggered enzymes can be able to introduce functional groups in an easily predictable way compared to plasma treatment. Enzymes considered for application in textile industry include mainly hydrolytic enzymes and oxidative enzymes.

In this chapter, recent promising results on enzymes for textile polymer surface treatment are reviewed and their application for polyester, PA, and PAN fiber surface treatment is discussed.

11.2 Enzymatic reactions

11.2.1 Polyester

Polyester is by far the most widely used polymer in synthetic textile manufacture. Within this large group of polymers, PET serves as component for more than 95% of all polyester textiles. Here, we focus our investigations on PET. The following Figure 11.1 shows the molecular structure of PET.

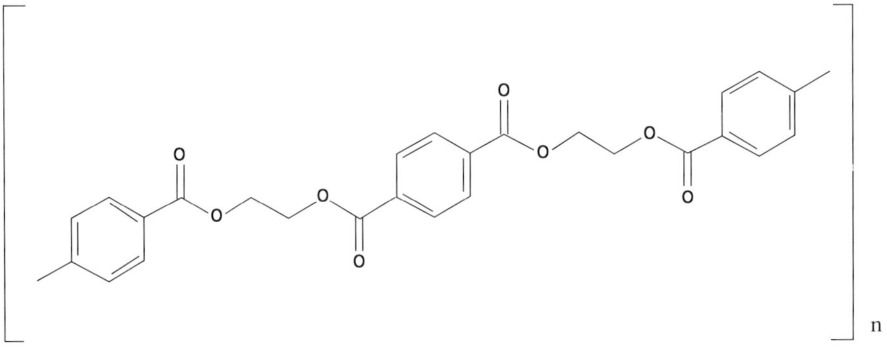

Figure 11.1. Molecular structure of polyethylene terephthalate (PET)

PET is known to be very hydrophobic and insoluble in water as well as in most organic solvents, which makes an enzymatic reaction at the surface of the polymer rather unlikely. Nevertheless, several enzymes have been found to carry out hydrolysis reactions at the PET surface [4]. Hydrolysis leads to an increase of free hydroxyl- and carboxylate end groups changing the surface properties of the treated material. This introduction of charged and functional groups directly leads to an increased hydrophilicity, which can be measured by determining water retention, rising height, or contact angles [5]. Furthermore, the increased amount of carboxylate and hydroxyl end groups facilitates the attachment of cationic dyes and could allow furthermore a reduction of coupling agents in special textile applications (e.g., PVC coating) [6].

11.2.1.1 Hydrolytic enzymes modifying PET

Lipases and esterases may potentially hydrolyze the ester bond of PET. Besides these hydrolases used in many industrial applications such as in detergents the potential of cutinases for PET modification has recently been assessed. Cutinases are enzymes that hydrolyze the plant cutin, which is the main component (between 40 and 80%) of the plant's cutical layer. Its monomers are oxygenated C_{16} hydroxyacids (e.g., 16-hydroxyhexadecanoic acid) and C_{18} hydroxyacids (e.g., 18-hydroxy-9,10-epoxyoctadecanoic acid) that form a polymeric network by ester linkages [7]. Cutinases belong to the family of serine hydrolases and they are specific for the hydrolysis of primary alcohol esters. Their substrate specificity is very broad, which can be seen from the wide range of chemical substances that can be hydrolyzed or synthesized [8]. Given the fact that cutinases act on insoluble esters of primary alcohols, it can be assumed that PET is a potential substrate.

Hydrolysis of PET with cutinases and other esterases has been proven as an increase of hydroxyl end groups of the polymer powder as well as on fabrics. A 48-h fabric treatment with an esterase increased the amount of hydroxyl end groups from 62 to 138 mmol/kg fabric [9]. Cutinases and so called polyesterases, which may be lipases or esterases, have been reported to reduce the pilling properties of PET [10, 11]. A pre-pilled PET fabric has been treated with a polyesterase from Genencor resulting in the removal of nearly all pills. A weight loss of about 4% correlates with the depilling effect [10]. *Hsieh and Cram* have tested the ability of lipases to hydrolyze the PET surface and achieved a water contact angle decrease from 78° (untreated fabric) to 40–60° depending on the lipase used. These results correlated with a change in water retention, which was found to be increased from 0.2 μL/mg fabric (untreated fabric) to 1.0 μL/mg fabric, showing strongly increased hydrophilicity [5]. Furthermore, PET hydrolyzing enzymes have been shown to improve color clarity in fabrics [11].

A vast number of esterases and lipases has been isolated and characterized and many enzymes are commercially available for manifold industrial applications. In contrast, less information is available about cutinases in the literature. In general, cutinases are very stable enzymes that can be produced from different sources in reasonable amounts and stabilized for long-term storage by solid formulations or in micellar liquids [12, 13]. Cutinases from fungi show considerable stabilities at acidic pH around 4 while cutinases from bacterial sources such as *Thermomonspora fusca* are reasonably stable up to pH 11 [8, 13]. Thermostability of cutinases has been reported to be as high as 70°C with a half-life time of 60 min at the pH optimum [13]. Thermostability of cutinases can even be increased by site directed mutagenesis. It has been described that an exchange of one or more negatively charged amino acids near the N-terminal of the protein by neutral or positively charged amino acids can increase the thermostability of the cutinase by at least 5°C [14]. Also, cutinases have been reported to be produced in high yields from carbon sources such as tomato peel or apple pomace, which are usually wastes or low value animal food [13]. These facts like good pH and thermostability and cheap production possibilities make cutinases applicable and competitive for industrial applications.

11.2.1.2 *Oxidative enzymes modifying PET*

Laccases are unspecific oxidoreductases that catalyze the removal of a hydrogen atom from the hydroxyl group of ortho- and para-substituted mono- and poly-phenolic substrates and from aromatic amines by one-electron abstraction. The substrate range of laccases can even be expanded by the use of electron mediators. Laccases have been reported to strongly enhance the hydrophilicity of a PET knitted fabric measured by determining the water contact angle and the rising height. Depending on the treatment time, the laccase activity in the assay and the mediator used, rising heights of 5.8 cm (laccase activity: 1000 nkat/g, treatment time: 300 s), 7 cm (laccase activity: 1000 nkat/g, mediator: violuric acid, treatment time: 300 s), and 7.3 cm (laccase activity: 1000 nkat/g, mediator: 2,2,6,6-tetramethylpiperidin-1-yloxy, treatment time: 300 s) compared to 0.2 cm for the untreated fabric have been observed [15]. However, the mechanism of these investigations has not been elucidated until now.

11.2.2 Polyamide

PAs used for textile purposes comprise Nylon® 6,6, which consists of adipic acid and bishexaneamine as monomers and Nylon 6 made of the monomer 6-aminocaproic acid (Figure 11.2) (laut kai beide wichtig).

Like PET, PA is quite recalcitrant to microbial degradation and therefore, only few enzymes are known with the potential to modify the polymer surface.

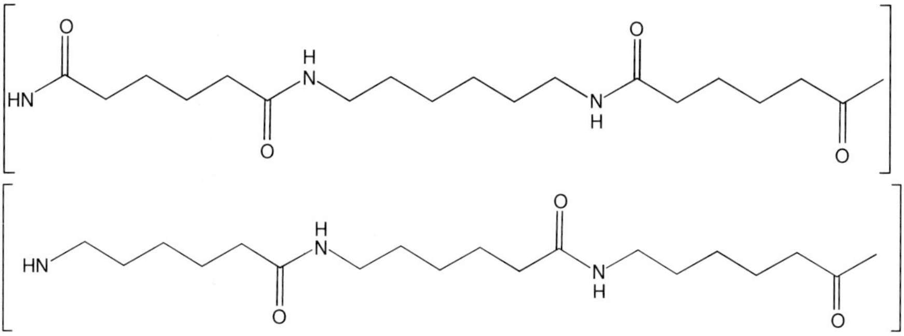

Figure 11.2. Molecular structures of Nylon® 6,6 (upper picture) and Nylon® 6 also known as Perlon®

The aim of PA modification is an increased hydrophilicity and an introduction of functional (e.g., ionic) groups for further chemical or enzymatic modification. In general, hydrolytic enzymes can act on the polymer surface resulting in an increased amount of free amino- and carboxylate end groups giving the PA similar properties as described for PET in chapter 2.1. Another possibility is the surface modification of PA by oxidative enzymes resulting in introduction of functional groups either by cleaving or retaining the backbone.

11.2.2.1 Hydrolytic enzymes modifying PA

It has been shown recently that several bacterial strains were found with the ability to use PA, namely Nylon® 6,6, and a model substrate for Nylon®, adipic acid bishexaneamide, as the only carbon source for growth. Preliminary experiments with enzyme preparations of the above mentioned bacterial strains showed a significant hydrolytic activity for adipic acid bishexaneamide [9]. These enzymes have been shown not to be inhibited by commercial protease inhibitors. Even though the nature and mechanism of the enzymes have not been revealed yet, they have potential to modify the PA polymer surfaces, as it has been shown for hydrolytic enzymes from *Comamonas acidovorans* [16].

11.2.2.2 Oxidative enzymes on PA

A white rod fungal strain IZO-154 has been isolated with the ability to degrade Nylon® 6,6 films during growth on glucose and ammonium tartrate. Investigations of changes in the molecular mass have shown a decrease indicating a cleavage at the backbone of the PA. Subsequent NMR analysis of the fibers revealed the emerging of four new end groups i.e., $-CHO$, $-NHCHO$, $-CH_3$, and $-CONH_2$ indicating an oxidative cleavage of C–C and non-amidic C–N bonds rather than hydrolytic cleavage at the amide bonds [17]. The isolation and characterization of the enzyme responsible for the Nylon® 6,6 degradation has been shown to be a manganese peroxidase although a different reaction mechanism has been suggested [18]. It can be assumed that the purified enzyme system can be used to modify PA fibers and fabrics, since the emerging end groups can change the PA surface significantly and may serve as anchor groups for other molecules (e.g., dyes, polymers, etc.). A strong deterioration can be avoided by shortening the treatment times. Deterioration is reported to take place after a 2-day incubation. Still, it has to be mentioned that manganese peroxidases are not reported to be very stable, and also their need for Mn(III) salts do not make them a competitive enzyme for industrial scales.

Other authors have isolated a *Phanerochaete chrysosporium* strain able to degrade Nylon® 6 fibers during growth on Nylon® 6 as the only nitrogen source. Degradation is reported to take place within several months of growth and has been followed by determination of the molecular weight calculated from relative viscosity. The molecular weight of the PA decreased from 16,900 g/mol to about 8500 g/mol [19]. Even though the mechanism of the biodegradation has not been elucidated until now, these preliminary investigations may result in the purification that modifies the PA surface oxidatively as it has been discussed by Deguchi et al. [17].

Fujisawa et al. have investigated a laccase-mediator system for its ability to degrade Nylon® 6,6. Using 1-hydroxybenzotriazole (HBT) as a laccase mediator, they have shown that Nylon® 6,6 membranes are disintegrated after a 2 days treatment [20]. Other investigations, where a laccase-mediator system was used to increase the hydrophilicity in Nylon® 6,6 have shown that the rising height as a parameter for hydrophilicity was enhanced from 1.8 (untreated) to 3.8 cm after a 300 s treatment with laccase and violuric acid as a mediator, and to 5.5 cm with laccase alone [15]. Despite these promising results the mechanism for laccase catalysed oxidation of polyamide remains to be elucidated. Laccases as potential enzymes for an industrial application show the advantage of being stable enzymes that can be produced very cheap in high amounts.

11.2.3 Polyacrylonitrile

PAN is a manmade fiber that is produced by radical polymerization of acrylonitrile. The following Figure 11.3 shows the molecular structure of PAN.

PAN usually has a molecular weight of about 55,000 g/mol and generally contains 5–10 mol% of a copolymer, e.g., vinyl acetate to disrupt the high crystallinity of the PAN polymer. Even though PAN is known to show high hydrophilicity and skin comfort compared to PET or PA, an improvement of these properties is still desired. Besides this, the moisture uptake and the dyeability with ionic dyes shall be improved by maintaining the good mechanical properties of PAN [9].

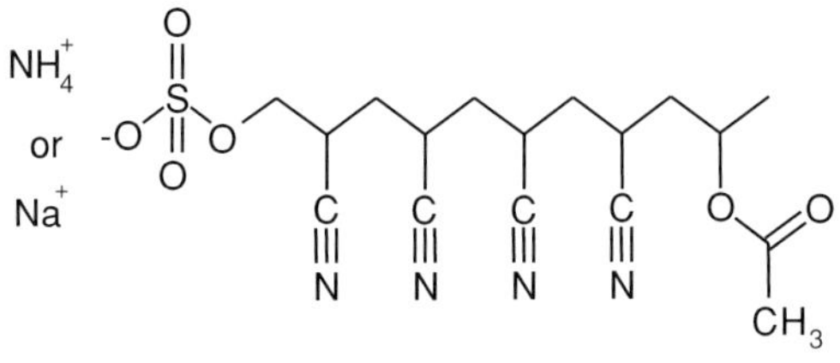

Figure 11.3. Molecular structure of polyacrylonitrile

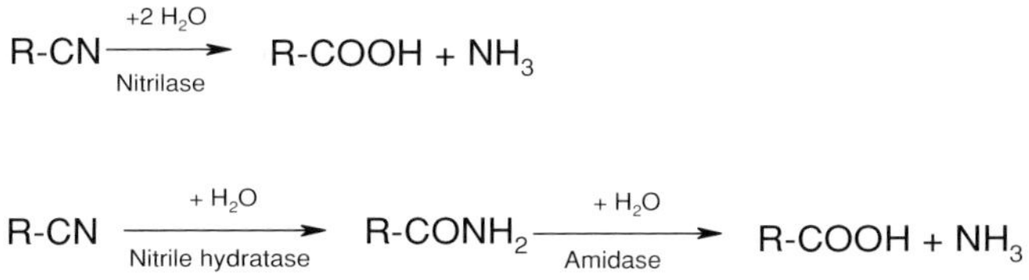

Figure 11.4. Reaction scheme of a nitrilase (upper picture) and a nitrile hydratases/amidase enzyme system

In general, nitrile hydrolyzing enzymes can be used to modify the PAN surface by hydrolyzing the nitrile groups to the corresponding amide or acid. Enzymes showing esterase activity may be used to modify the vinyl acetate groups in the PAN.

Within the large group of nitrile hydrolyzing enzymes, two pathways of nitrile hydrolysis can be described as illustrated in the following Figure 11.4.

As it can be seen from the picture, nitrilases are enzymes that catalyze the hydrolysis of a nitrile directly to the corresponding acid, most probably forming an acylenzyme as reaction intermediate [21]. Nitrile hydratase/amidase enzyme systems catalyze the hydrolysis in a two-step reaction [22]. Both enzymatic ways to change the PAN surface are possible; nevertheless, the nitrile hydratase/amidase enzyme system has been reported more often to be used for modification of the PAN surface.

Tauber et al. have isolated a nitrile hydratase and amidase from a *Rhodococcus rhodochrous* strain able to modify the surface of PAN. The hydrolysis of the nitrile groups to the corresponding acid was observed by determination of the resulting NH_3 and by XPS analysis. The resulting amino groups were measured by determining the increase of the K/S values with Methylene blue and Coomassie brilliant blue. Treatment of PAN fabric with the nitrile hydratase/amidase enzyme system resulted in a significant increase of amide groups but no corresponding acid groups. The modified PAN showed much higher K/S values after Methylene blue and Coomassie brilliant blue staining [23].

Treatment of PAN with nitrile hydratases from *Brevibacterium imperiale* and from *Corynebacterium nitrilophilus* resulted in an increase of amide groups at the PAN surface. This change was monitored by XPS surface analysis and lead to properties like increased hydrophilicity, measured by the contact angle method, and dyeability with acid dyes coupling to the protonated amide nitrogen atoms [24].

11.3 Conclusion

Since the middle of the 1990s, investigations on the surface modification of synthetic fibers have been made. Shortcomings in the technologies used for textile processing such as use of strong and sometimes hazardous chemicals, insufficient quality of the fibers and fabrics, and the need for investigations in technologies for special applications have resulted in the assessment of the potential of enzymes for surface modification of synthetic textiles. Using the ability of microorganisms for fast adaptation to new substrates (e.g., synthetic polymers) and the knowledge of enzyme engineering and genetics, promising results have been achieved until now.

For PET, PA, and PAN, the hydrophilicity of the fibers/fabrics have been proven to increase due to enzymatic action on the polymer surface. Furthermore, treatment with hydrolytic enzymes promise better dyeability with ionic dyes and the newly introduced end groups give space for further reactions (such as adsorptive bonding of special agents such as flame retardants or other finishers). An introduction of functional groups into PET and PA without hydrolysis of the polymer backbone by oxidative enzymes has also been discussed.

The enzymatic surface modification of manmade polymers is a promising field of research giving rise to the hope for new competitive and environmentally friendly technologies.

References

1. Engelhardt, A. (2003). The Fiber Year 2002, The Saurer Group, Winterthur, Switzerland.
2. Hochard, F.; Levalois-Mitjaville, J.; De Jaeger, R.; Gengembre, L.; Grimblot, J. Plasma surface treatment of polyacrylonitrile films by fluorocarbon compounds. *Appl. Surface Sci.* **1999**, *142*, 574–578.
3. Tusek, L.; Nitschke, M.; Werner, C.; Stana-Kleinscheck, K.; Ribitsch, V. Surface characterization of NH₃ plasma treated polyamide 6 foils. *Colloids Surfaces A: Physicochem. Eng. Aspects* **2001**, *198*, 81–95.
4. Deckwer, W.; Mueller, R.; Kleeberg, I.; Van Den Heuvel, J. Ester cleaving enzyme of *Thermomonospora fusca* and its use in degradation of polyesters. *PCT Int. App.* **2001**, WO 2000-EP7115.
5. Hsieh, Y.; Cram, L. Enzymatic hydrolysis to improve wetting and absorbency of polyester fabrics. *Tex. Res. J.* **1998**, *68*(5), 311–319.

6. Gübitz, G.; Cavaco-Paulo, A. New substrates for reliable enzymes: Enzymatic modification of polymers. *Curr. Opin. Biotechnol.* **2003**, *14*, 577–582.

7. Heredia, A. Biophysical and biochemical characteristics of cutin, a plant barrier biopolymer. *Biochim. Biophys. Acta* **2003**, *1620*, 1–7.

8. Carvalho, C.; Aires-Barros, M.; Cabral, J. Cutinase structure, function and biocatalytic applications. *El. J. Biotechnol.* **1998**, *1*(3), 160–173.

9. Fischer-Colbrie, G.; Heumann, S.; Liebminger, S.; Almansa, E.; Cavaco-Paulo, A.; Gübitz, G. M. New enzymes with potential for PET surface modification. *Biocat. Biotrans.* **2004**, *22*, 341–346.

10. Yoon, M.; Kellis, J.; Poulose, A. Enzymatic modification of polyester. *AATCC Rev.* **2002**, *2*, 312–318.

11. Andersen, B.; Borch, K.; Abo, M.; Damgaard, B. A method of treating polyester fabrics. *PCT. Int. App.* **1998**, WO 1999/001604.

12. Gouda, M.; Kleeberg, I.; Van Den Heuvel, J.; Mueller, R.; Deckwer, W. Production of a polyester degrading extracellular hydrolase from *Thermomonospora fusca*. *Biotechnol. Progr.* **2002**, *18*, 927–934.

13. Fett, W.; Wijey, C.; Moreau, R.; Osman, S. Production of cutinase by *Thermomonospora fusca* ATCC 27730. *J. Appl. Microbiol. Biotechnol.* **1999**, *86*, 561–568.

14. Abo, M.; Fukuyama, S.; Matsui, T. Cutinase variants. *Int. PCT. App.* **2000**, WO 00/34450.

15. Miettinen-Oinonen, A.; Puolakka, A.; Nousiainen, P.; Buchert, J. Modification of textile fibres by laccase. COST 847 Workshop, January 29–30, 2004, Belfast, UK.

16. Crouzet, J.; Faucher, D.; Favre-Bulle, O.; Jourdat, C.; Petre, D.; Pierrard, J.; Thibaut, D.; Guitton, C. Enzymes and micro-organisms with amidase activity which hydrolyse polyamides. *Int. PCT. App.* **1997**, WO 97/04083.

17. Deguchi, T.; Kakezawa, M.; Nishida, T. Nylon biodegradation by lignin-degrading fungi. *Appl. Environ. Microbiol.* **1997**, *63*(1), 329–331.

18. Deguchi, T.; Kitaoka, Y.; Kakezawa, M.; Nishida, T. Purification and characterization of a nylon-degrading enzyme. *Appl. Environ. Microbiol.* **1998**, *64*(4), 1366–1371.

19. Klun, U.; Friedrich, J.; Krzan, A. Polyamide-6 fibre degradation by a lignolytic fungus. *Polym. Degrad. Stab.* **2003**, *79*, 99–104.

20. Fujisawa, M.; Hirai, H.; Nishida, T. Degradation of polyethylene and Nylon-66 by the laccase-mediator system. *J. Polym. Environ.* **2001**, *9*(3), 103–108.

21. Brenner, C. Catalysis in the nitrilase superfamily. *Curr. Opin. Struct. Biol.* **2002**, *12*, 775–782.

22. Kobayshi, M.; Shimizu, S. Nitrile hydrolases. *Curr. Opin. Chem. Biol.* **2000**, *4*, 95–102.

23. Tauber, M.; Cavaco Paulo, A.; Robra, K.; Gübitz, G. Nitrile hydrolase and amidase from *Rhodococcus rhodochrous* hydrolyze acrylic fibers and granular polyacrylonitriles. *Appl. Environ. Microbiol.* **2000**, *66*(4), 1634–1638.

24. Battilstel, E.; Morra, M.; Marinetti, M. Enzymatic surface modification of acrylonitrile fibers. *Appl. Surface Sci.* **2001**, *177*, 32–14.

Chapter 12

ENZYMATIC MODIFICATION OF FIBERS FOR TEXTILE AND FOREST PRODUCTS INDUSTRIES

William Kenealy[1], Gisela Buschle-Diller[2], and Xuehong Ren[2]

[1] *U.S. Department of Agriculture, Forest Service, Forest Products Laboratory, Madison, Wisconsin, U.S.A.*
[2] *Textile Engineering Department, Auburn University, Auburn, Alabama, U.S.A.*

Abstract

A variety of enzymes are available for the surface modification of cellulosic fibers, both in the area of textile applications and for pulp and paper applications. Enzymatic treatment conditions are milder, less damaging for the fiber, and are environmentally friendly while producing effects comparable to chemical treatments. Surface modifications can be achieved by oxidative and/or hydrolytic enzymes. Some of the enzymatic processes have recently attained commercial importance and more systems are being developed. The following chapter will review current research in the application of oxidoreductases and hydrolases that are valuable for textile and forest products industries.

Keywords: surface modification; oxidoreductase; laccase; peroxidase; hydrolase; cellulose-binding domain.

12.1 Introduction

Fibers derived from plants have a surface chemistry that is inherent in the structure and the source of the material. These fibers will often have a set of properties, such as water-binding capability, flexibility, rigidity, hydrophilic and hydrophobic regions, and the ability to adhere to themselves and other materials, which is dependent on the structure and assembly of the major components of the fiber (hemicellulose, cellulose, and lignin). For economic or supply availability concerns, it is often desirable to modify these properties, thus altering

J. V. Edwards et al. (eds.), Modified Fibers with Medical and Specialty Applications, 191–208.

the fiber or its surface, to suit the end-product. A variety of chemical techniques have been developed to achieve this goal.

One way of changing fiber surface properties is to remove components of the fiber. The first steps in both the textile and the pulp and paper industries are usually processes to separate the cellulosic material from non-cellulosic impurities. The well-known processes of kraft or sulfite pulping result in a cellulosic fiber with good paper strength and little lignin or hemicellulose content. The fiber is bludgeoned into the suitable form for the product. In the textile industry, scouring and bleaching processes remove oils, fats, pectin, hemicellulose, and coloring matter and render the fiber material clean, water absorbent, and prepared for further modification, such as dyeing.

Another possibility to alter a fiber is to add components. Pulps with high cellulosic content, for example, can be made into cellulose acetate films and carboxymethylcellulose derivatives depending on whether the carboxylic acid of the added acetyl group is involved in an ester or is free to react in an aqueous solution. The result of these additions is a more soluble fiber in the case of carboxymethylcellulose or a more insoluble fiber in the case of cellulose acetate. Acetyl esters of wood created with acetic anhydride can make the wood more dense and more hydrophobic [1]. Attachment of carboxylic acids also affects the strength of paper. The presence of carboxylic acids [2–4] and the effect of different cations [5–7] bound to the dissociated acid have been shown to have significant effects on the strength of paper. Cellulosic fibers for textile uses can be modified by addition of hydrophobic groups for water repellency, cross-linking agents for improved performance, or softeners for enhanced hand, for example [8].

A third method of fiber alteration is to change the nature of fiber functional groups; for instance, by oxidation reactions to increase the acidic groups available for better bonding in the case of pulp fibers [2]. A variety of bleaching techniques can be used to remove color from pulps used for writing papers by either oxidative or reductive methods [9–14]. As a result of bleaching, the chromophore is removed, altered, or destroyed, although in some cases brightness reversion may occur [10, 15, 16].

Chemical treatments of fiber involving harsh reaction conditions or highly reactive chemicals create problems with non-specific reactions, potential hazards to the user, and cost in yield of the products or loss of desirable components. Such conditions can cause a reduction in degree of polymerization, loss of hemicellulose, and oxidation reactions at end-groups so that the fiber negatively changes its properties and loses strength. It is of utmost importance to maintain mild conditions for modification of cellulosic surfaces when the fiber integrity is to be maintained [17]. As a consequence of these limitations of chemical systems and the nature of the fiber, the use of enzymes to perform some of these reactions has been investigated [18–21].

Enzymes, applied at mild conditions, have the advantage of being specific to their substrates and more easily controlled, and they are viewed as environmentally friendly and less damaging to the fiber [18, 20, 22]. The disadvantages with enzymes are that they are often not robust enough for the process [23]; they are expensive, especially when used in a single step without enzyme recycle, and frequently, they are not available in commercially useful quantities [18].

Enzymatic surface modifications of fibers can be created by oxidative and/or hydrolytic enzymes. Oxidoreductases, such as peroxidases, laccase, and cellobiose dehydrogenase, all have uses in the modification of fiber surfaces [10, 18–21]. Likewise hydrolases, such as cellulases, hemicellulases, pectinases, amylases, lipases, and proteases, are able to modify fiber surfaces [18–20, 22]. Attachment of various functional groups or enzymes by the use of cellulose-binding domains (CBDs) is also becoming more common [24, 25]. We will review recent research on enzymatic modifications of fibers for the textile and forest product industries, including results from experiments performed in our laboratories.

12.2 Oxidative enzymes and their applications

12.2.1 Laccases for pulp and paper modification

Laccases are enzymes that catalyze the oxidation of a variety of phenolic and similar compounds [26, 27] and are implicated in the degradation of lignin [28]. Laccases have been identified in both fungi and bacteria [29]. The enzymes abstract an electron from substrates containing phenolic or aromatic nitrogen, which produces a free radical, and reduce oxygen to water, as shown in Figure 12.1 for hydroquinone. Laccases can degrade model lignin compounds [30–32] and have been useful in the removal of lignin from pulps [33]. Removing lignin is important for the properties of chemical pulps (improved flexibility, better fiber to fiber bonding, increased sheet strength) and the brightness of the paper. In addition to delignification, laccases are able to bleach, oxidize, and graft materials onto the fiber surface [19, 28, 34].

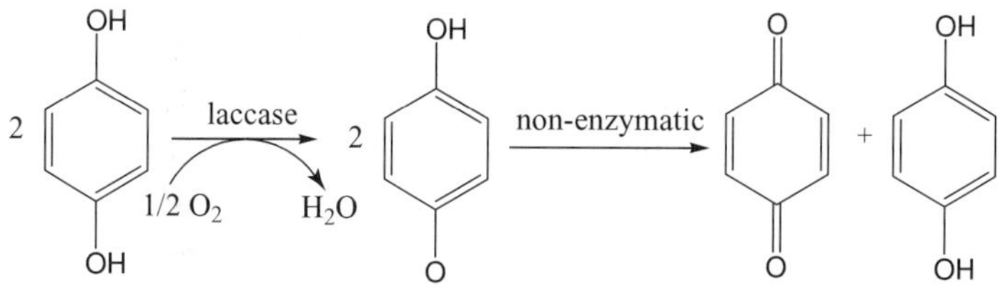

Figure 12.1. Enzymatic abstraction of electron from hydroquinone

Delignification of pulps and various mediator combinations with laccases have been studied by many authors [19, 28, 35–41]. Delignification is most prominent when the lignin content of the pulp is high [33]. For effective delignification, laccase requires the addition of a mediator [42]. The mediator oxidizes lignin at a distance from the enzyme and also reacts with compounds that are not substrates for laccase. As depicted for hydroquinone (Figure 12.1), laccase abstracts an electron from a suitable substrate with the reduction of molecular oxygen. The enzymatic free radical products may undergo further oxidative and non-enzymatic reactions, including polymerization and dismutation [41, 43].

The reaction and decay of the free radicals generated in a complex substrate such as lignin are difficult to foresee, and the products of the reaction are complex. The complexity of the products of the laccase reaction depends in part on the resonance structures of the free radical, the solubility of the product, other compounds available to react with the radical, and the longevity of the radical [41, 44]. Despite this complexity, carefully designed reactions can predict the reactions taking place and obtain product in reasonable yield [45].

Mediators are substrates of laccase able to react with structures that the laccase enzyme cannot oxidize [30, 46]. Mediators can be long-lived free radicals or transition metal complexes [38, 47]. They can be natural products of fungi or other chemicals [48]. The use of laccase in bleaching and delignification requires mediators, and commercial application is hampered by the lack of an inexpensive source [42, 49]. New mediators and laccase substrates are being sought and reported [27, 41].

The laccase mediator for bleaching may have different effects depending on the material being bleached or delignified [50–53], the enzyme used [27], and the mediator employed [46]. The mechanism of laccase mediator reactions has been investigated—some proceed by a radical hydrogen atom transfer (using mediators, 1-hydroxybenzotriazole, violuric acid, or hydroxyphthalimide) while the most often reported mediator, 2,2′-azino-*bis*(3-ethlybenzthiazoline-6-sulfonic acid), can proceed via an electron transfer mechanism [43].

The reaction of laccase on fiber can create a variety of effects. Wood-based products can be oxidized, providing better strength in composites [54], papers and fiberboards [55–57], and reducing energy costs [58]. For example, reactions of laccase with radiata pine chips resulted in energy savings during mechanical refining and a stronger handsheet from the pulp [58]. Laccase treatment was shown to increase the strength of various fiberboards [59, 60] and paper [55].

A relatively new area for laccase application is the grafting of materials to lignocellulose [61, 62]. The laccase reaction with lignin builds up a phenoxy-radical charge in the material [62]. The radicals in these polymers appear to be either long-lived or are rapidly regenerated by the enzyme. These radicals have been used to graft various materials onto the fiber, including carbohydrates [62], acrylates [63, 64], and acrylamides [65, 66]. Other polymeric materials [54, 62] can also be attached, thus modifying the fiber surface. The addition of polymers changes the material properties of the cellulosic material, increasing its strength or water-holding capacity.

Exploration of the vast substrate range of laccase and the effects of different substrates on a lignocellulose product are in its infancy [67]. Phenolic acids can be attached to pulps using laccase and various substrates such as gallic acid [68], 4-hydroxybenzoic acid [69], and 4-hydroxyphenylacetic acid [67, 70]. The attachment of the phenolic acids increases the burst and tensile strength of test handsheets, presumably by the presence of the organic acid. However, the attachment of phenolic materials that alter the surface characteristics of fiber, such as resorcinol, phloroglucinol, and catechol, does not increase the strength of handsheets [67]. The effect of laccase treatment of substrates must be empirically derived because of reactions with oxygen and other radicals generated in the reaction, polymerization of the substrate, and condensation onto lignocellulose. The nature of the laccase substrate can only suggest that a particular effect on the surface chemistry may be obtained.

12.2.2 Textile fiber applications of laccases

Laccases have also found applications in textile processes. The introduction of a more user-friendly laccase formulation to the textile-finishing market spurred an interest in its use for treatment of indigo-dyed cotton denim. Several reports of laccase bleaching appeared in the literature by the turn of the last century [71]. Laccase was applied together with a suitable mediator to create a bleached-out look to jeans, which was fashionable at that time. In this process, the indigo chromophore was transformed into isatin [71], and backstaining was reduced or avoided. This process worked especially well for indigo while laccase was found to be ineffective for bleaching undyed cotton. Tzanov et al. [72] developed a laccase pre-treatment that, when followed by a traditional hydrogen peroxide bleach, resulted in a whiter cotton fabric than possible without the pretreatment.

Lignin probably plays an important role in the reaction mechanism of laccase bleaching. Raw cotton probably could not be satisfactorily bleached using only laccases because cotton has little lignin. Ossala and Galante [73] published a comparative study on scouring of flax rove with a variety of enzymes. They found, however, that laccase performed the least effective of all investigated enzymes. Hydrolases, such as pectinase, gave superior results.

Similar to the case with the pulp and paper industry, laccases have been tested in reactions to both bleach and graft materials to textiles. Shin and

Cavaco-Paulo [74] made use of the laccase reaction by *in situ* dyeing of wool fibers with laccase and small phenols as substrates. Similar reactions were patented by Aaslyng et al. [75] for human hair.

12.2.3 Peroxidase applications for pulp

Manganese peroxidase, lignin peroxides, and other peroxidases have also been used for pulp-bleaching reactions [18]. The use of peroxidases requires the addition of hydrogen peroxide, which at high concentrations also inhibits the enzyme. One approach is to meter in the hydrogen peroxide at low concentrations, matching the consumption rate with the supply rate. Another is to use an enzyme like glucose oxidase or another hydrogen peroxide generating system at levels that produce the needed amount of substrate. The generation or metering of hydrogen peroxide alleviates the inhibition of the enzyme.

The requirement of an additional substrate such as manganese is a small problem for the marketable use of these enzymes compared to the lack of commercially available amounts of the enzymes. Laccases, even with the requirement for mediators, have been favored since they have been relatively easy to produce in heterologous expression systems. Peroxidases have long been available, but the use of manganese peroxidase and lignin peroxidase has been limited by the lack of commercial quantities of these enzymes.

Research on the use of manganese peroxidase has indicated a variety of potential applications. Manganese peroxidase has been used for the bleaching of chemical pulps [76], delignification of pulps [77], and treatment of pulps to lessen electrical refiner energy and improve handsheet strength [78]. Lignin peroxidase is promising since it can react with non-phenolic components of lignin and does not require the addition of manganese. The commercialization of processes using lignin peroxidase has been hampered by the recalcitrant heterologous expression of the enzyme.

Textile applications of peroxidases for bleaching of cotton and of lignin-containing fibers, such as linen, are being explored, with limited success [79]. Possible options are combining compatible oxidoreductases or applying the enzymes in suitable sequences of optimum pH and temperature. For example, the application of manganese peroxidase in the first step followed by glucose oxidase results in slightly higher whiteness levels than the application of glucose oxidase alone. Such process modifications might have a greater potential than does treatment with only one peroxidase. However, the process costs are still high and might not be economically justified at present.

Cellobiose dehydrogenase has also been explored to modify the structure of lignocellulosic materials. While under certain conditions cellobiose dehydrogenase can catalyze lignin degradation [80], the cost and action of this enzyme will probably prohibit commercial application in pulp bleaching and delignification [81]. The modification of carbohydrate by creating carboxylic acids is a potential cost-effective application of cellobiose dehydrogenase [81].

12.2.4 Glucose oxidase bleaching of textile substrates

Flavo-enzymes, such as glucose oxidase, have long been regarded as uneconomical and ineffective for textile applications except for their incorporation in laundry formulations [82] for stain removal and as anti-redeposition agents [83].

Glucose oxidase specifically oxidizes β-D-glucose to δ-D-gluconolactone, releasing hydrogen peroxide, which then can be used for bleaching of cellulosic materials. A closed-loop process using glucose-rich effluents from enzymatic desizing and scouring in conjunction with bleaching whitened cotton to a level close to that of chemically bleached material [84]. A similar bleaching process was established using glucose oxidase immobilized on alumina or glass support, which increased the stability of the enzymes and offers the possibility of enzyme recycling to reduce the cost of the process [85].

We performed bleaching studies involving the treatment of unbleached linen fabric with laccase alone and in combination with glucose oxidase. Laccase was first applied as the sole enzyme with minimal gain in whiteness. Adding glucose oxidase in a second step and finally, in the third step, raising the pH and temperature to 10.5 and 90 °C, respectively, resulted in whiteness levels nearly comparable to that of chemically bleached linen [79, 84].

12.3 Hydrolases

12.3.1 Xylanase for pulp and paper applications

Oxidative enzymes are not the only useful enzymes in the modification of pulp and paper. Perhaps the most successful enzymatic application in the pulp and paper industry has been the use of xylanase to assist in bleaching of kraft pulps. Xylanases have been shown to aid in the bleaching of many different softwoods and hardwoods [18, 86]. Both chemical and mechanical pulps have benefited by xylanase treatment [18]. Since the first description of the use of xylanase to boost the bleaching of pulps [87], there has been an acceleration in the identification of new xylanases and continuing research into developing xylanases that are more suitable for the elevated temperatures, alkaline pH, and other conditions (presence of proteases) in the stages where the enzyme is used [18, 86, 88–90].

In bleaching, xylanase acts by cleaving the xylan, producing shorter, easier to remove oligosaccharides, which aids in the removal of colored compounds [91–94] and bleach-consuming chemicals such as hexenuronic acid [95–98]. In addition, the removal of xylan enhances the removal of lignin, which is also a major contributor to color in the pulp. Synergistic action was noted when different xylanases were used [99] or a combination of xylanase and oxidative enzymes [100]. Xylanase treatment decreases the amount of bleaching chemicals required to attain the desired brightness. Again, as with laccase, several reviews have been published on the use of xylanases in enhancing pulp bleaching [18, 19, 101].

Other hydrolases have also been reported as suitable for the bleaching of paper pulps. Cellulases have been useful in the removing colored material from recycled yellow pages and print in recycled pulps [102]. Mannanases and xylanases can act synergistically to aid in bleaching some pulps [103–109]. These enzymes do not work on all pulps since there are considerable variations in the type of pulp (mechanical, thermomechanical, kraft, sulfite, or recycled fiber) as well as in the tree species used in the pulping process [18]. Different enzymes will often work better on a specific pulp, so the appropriate enzyme and dose are determined by empirical testing.

Xylanases have also been included in the enzymatic scouring formulation for raw cotton to reduce the amount of seed coat fragments. Immature cotton fibers and seed coat fragments are problems for the textile industry because they cannot be dyed, especially with dyes applied at neutral pH, and thus remain as small spots. Seed coat fragments and immature cotton fibers, which are low in cellulose, can be removed to a great extent by a harsh chemical scouring process under alkaline conditions. Enzymatic scouring with pectinases and cellulases, however, is specific for pectins and cellulose, respectively, thus ineffective on both seed coat fragments and immature cotton. The addition of xylanases has been helpful to some extent [79].

Lipases are used in both the pulp and paper and textile industries. Lipases remove pitch from pulps [20, 110]. Pitch deposits are imperfections in the paper caused by fats and resins. These deposits can accumulate and cause problems during the papermaking process and limit the value of the paper. Lipase treatment of the pulp cleaves the triglycerides and allows the fatty acids to be removed with other waste in the rinse water.

Cellulases have been applied in the pulp and paper industry with great caution [111]. The degradation of the cellulosic fibers must be avoided, but enzymes can be used to remove the cellulose that does not provide value to the product. Cellulases have been particularly effective in improving the drainage [102, 112–114] and deinking of recycled fiber [115–120] as well as cellulose fabrics [121]. Cellulases have also been used successfully followed by peroxide bleaching for low-quality recovered paper containing unbleached and mechanical fibers which are a major obstacle in recycling lower quality paper mixes [122]. Hemicellulases, lipases, and esterases have also been employed in deinking of recycled fiber [119].

Cellulase-assisted deinking can involve the use of other additives such as surfactants [123–125]. If floatation is the primary method of ink removal, the cellulase application must be tested to determine if removal of the amorphous cellulose on ink particles will be advantageous or detrimental to the removal of the specific type of ink used [126]. The removal of the amorphous regions can improve the drainage of the recycled fiber and has been used both before and after refining [19]. Removal of stickies contaminants with cellulase treatment during the recycling process not only enhanced flotation removal efficiency, but also allowed the process to be conducted at neutral pH. The enzyme process resulted in a substantially cleaner process overall [127].

Much research has been published on using enzymes to modify the properties of textiles. Lipases, amylases, and proteases have been included in a variety of detergent formulations to aid in removing stains [128–130]. Proteases have been used for shrink-proofing and softening wool fibers and for degumming silk [131, 132]. Cellulases are now commercially applied to biopolish and manually improve cotton [133–138] and to achieve the "peach-skin" effect on lyocell fibers by fibrillation [139, 140]. The macroscopic effect of cellulases is the removal of small protruding fibrils from the fiber surface with or without mechanical impact, thus enhancing softness and color brilliance by an "abrasive, bio-polishing" action.

Cotton is usually scoured in an alkaline solution. The process requires a considerable amount of water and generates waste that must be neutralized and disposed. Bioscouring using enzymes such as pectinases or a mixture of cellulases, pectinases, and xylanases can be done at neutral pH [141–146]. Because of the specificity of the enzymes, the fiber damage is minimal. Additionally, less waste water is generated. The end result is a stronger material with good absorbency.

12.3.2 CBDs as means of attachment

One new application is to establish the binding of various enzymes and other materials to cellulosic fibers using the CBDs of fungi and bacteria [147]. The fungal CBD is relatively short and can be added to the coding sequence of many enzymes (both carboxy- and amino-terminal extensions) to provide a functional domain able to fold properly and bind to cellulose [24, 25]. Bacterial CBDs are also used, and expression cassettes are available from molecular biology supply companies; the CBDs from both bacteria and fungi can be incorporated into the sequence of a protein to be expressed [148].

The simple binding of CBDs is sufficient to alter the dye affinity of cotton [149] and the properties of the pulps to which it is attached [150]. The addition of CBD increases the strength and drainability of secondary fibers [151]. As measured by atomic force microscopy, the binding of a CBD decreases repulsion between cellulose surfaces [152]. If the CBD is multimeric, there is also a greater adhesion between cellulose surfaces [152, 153]. CBD attachment alters the appearance of ramie cotton fibers, greatly smoothing their appearance.

The binding of different functional groups to cellulose has also been explored. Combining a CBD with a metal chelating peptide produces a product with the ability to remove metals [154]. Attaching an indicator enzyme to a CBD provides a solid-based device to detect the presence of a variety of components and is used in diagnostic and research materials [155]. Different CBDs bind to different regions of cellulose and with varying affinities. The presence of more than one cellulose-binding module per protein can influence the avidity with which the protein binds to cellulose.

Natural polymers such as xyloglucan can also be used to bind to cellulose and provide new chemistries on the surface of cellulosic fibers. Modified xyloglucan oligosaccharides have been incorporated into xyloglucans using xyloglucan endoglycosylase [156]. The affinity of xyloglucans for cellulose fibrils has made them useful in both textiles [156] and papermaking [157]. The alteration of the chemistry present in the xyloglucan will further the use of the xyloglucans in modifying cellulosic materials.

12.4 Conclusions

Enzymes have provided new products and have improved the quality and economics of processes in both the textile and forest products industries. In a capital-intensive industry like pulp and paper, there is a reluctance to use new processes if they require additional capital investment. The process must be able to fit the existing equipment. The techniques of molecular biology and heterologous expression allow the adaptation of enzymes to the process used. Even with these possibilities, the cost of enzymes is often too high for the action of the enzyme. New enzymes are being identified by basic research and genomics studies. The application of new enzymes can be expected when applied research tests their suitability for given applications.

References

1. Rowell, R., Chemical modification of wood. *Forest Prod. Abstr.* **1983**, *6*(12), 363–382.
2. Barzyk, D.; Page, D. H.; Ragauskas, A. Acidic group topochemistry and fibre-to-fibre specific bond strength. *J. Pulp Paper Sci.* **1997**, *23*(2), J59–J61.
3. Scallan, A. M. The effect of acidic groups on the swelling of pulps: A review. *Tappi J.* **1983**, *66*, 73–75.
4. Katz, S.; Liebergott, N.; Scallan, A. M. A mechanism for the alkali strengthening of mechanical pulps. *Tappi J.* **1981**, *64*(7), 97–100.
5. Scallan, A. M.; Grignon, J. The effect of cations on pulp and paper properties. *Sven. Papperstidning* **1979**, *82*(2), 40–47.
6. Lindstrom, T.; Carlsson, G. The effect of chemical environment on fiber swelling. *Sven. Papperstidning* **1982**, *85*(3), R14–R20.
7. Salmen, L. Influence of ionic groups and their counterions on the softening properties of wood materials. *JPPS* **1995**, *21*, J310–J315.
8. Slate, P. E. *Handbook of Fiber Finish Technology* (ed.). Marcel Dekker, New York, **1998**, 319–447.
9. Bucher, W. Bleaching 101: The basics of bleaching. *Solutions!* **2004**, *87*(February), 36–37.
10. Wong, K. K. Y.; Mansfield, S. D. Enzymatic processing for pulp and paper manufacturing. *Appita J.* **1999**, *52*(6), 409–418.
11. Johnson, A. P. Fitting together the ecf–tcf jigsaw. *Appita J.* **1994**, *47*(3), 243–250.
12. McDonough, T. J. Recent advances in bleached chemical pulp manufacturing technology. *Tappi J.* **1995**, *78*(3), 55–62.
13. Suchy, M.; Argyropoulos, D. S. Catalysis and activation of oxygen and peroxide delignification of chemical pulps: A review. *Tappi J.* **2002**, *1*(2), 1–18.
14. Li, K.; Collins, R.; Eriksson, K. E. Removal of dyes from recycled paper. *Prog. Paper Recycling* **2000**, *10*(November), 37–43.
15. Buchert, J.; Bergnor, E.; Lindblad, G.; Viikari, L.; Ek, M. Significance of xylan and glucomannan in the brightness of kraft pulp. *Tappi J.* **1997**, *80*(6), 165–171.
16. Forsskahl, I.; Tylli, H.; Olkkonen, C. Participation of carbohydrate-derived chromophores in the yellowing of high-yield and tcf pulps. *J. Pulp Paper Sci.* **2000**, *26*(7), 245–249.
17. Baiardo, M.; Frisoni, G.; Scandola, M.; Licciardello, A. Surface chemical modification of natural cellulose fibers. *J. Appl. Polym. Sci.* **2002**, *83*, 38–45.
18. Kenealy, W. R.; Jeffries, T. W. Enzyme processes for pulp and paper: A review of recent developments. In: Goodell, B.; Nicholas, D.D.; Schultz, T. P. (Eds.) *Wood Deterioration and Preservation: Advances in Our Changing World.* American Chemical Society, Washington, **2003**, 210–239.
19. Bajpai, P. Application of enzymes in the pulp and paper industry. *Biotechnol. Prog.* **1999**, *15*(2), 147–157.
20. Kirk, T. K.; Jeffries, T. W. Roles for microbial enzyme in pulp and paper. In: Jeffries, T. W.; Viikari, L. (Eds.) *Enzymes for Pulp and Paper Processing.* American Chemical Society, Washington, **1996**; Symposium series 655, 1–14.
21. Eriksson, K. E. Biotechnology in the pulp and paper industry: An overview. In: Eriksson, K. E.; Cavaco-Paulo, A. (Eds.) *Enzyme Applications in Fiber Processing.* American Chemical Society, **1998**; Symposium Series 687, 2–14.
22. Daniels, M. J. Using biological enzymes in papermaking. *Paper Technol.* **1992** *33*(June), 14–17.
23. Klibanov, A. Improving enzymes by using them in organic solvents. *Nature Biotechnol.* **2001**, *409*, 241–246.

24. Tomme, P.; Warren, A. J.; Miller, R. C. J.; Kilburn, D. G.; Gilkes, N. R. Cellulose-binding domains: Classification and properties. In: Saddler, J. N.; Penner, M. H. (Eds.) *Enzymatic Degradation of Insoluble Carbohydrates*. American Chemical Society, Washington **1995**; ACS Symposium series 618, 142–163.

25. Tomme, P.; Boraston, A.; McLean, B. W.; Kormos, J.; Creagh, A. L.; Sturch, K.; Gilkes, N. R.; Haynes, C. A.; Warren, A. J.; Kilburn, D. G. Characterization and affinity applications of cellulose-binding domains. *J. Chromatogr. B* **1998**, *715*, 283–296.

26. Xu, F. Oxidation of phenols, anilines, and benzenethiols by fungal laccases: Correlation between activity and redox potentials as well as halide inhibition. *Biochemistry* **1996**, *35*, 7608–7614.

27. Xu, F.; Kulys, J. J.; Duke, K.; Li, K.; Krikstopaitis, K.; Deussen, H. J. W.; Abbate, E.; Galinyte, V.; Schneider, P. Redox chemistry in laccase-catalyzed oxidation of *n*-hydroxy compounds. *Appl. Environ. Microbiol.* **2000**, *66*(5), 2052–2056.

28. Yaropolov, A. I.; Skorobogat, K. O. V.; Vartanov, S. S.; Varfolomeyev, S. D. Laccase: Properties, catalytic mechanism, and applicability. *Appl. Biochem. Biotechnol.* **1994**, *49*(3), 257–280.

29. Martins, L. O.; Soares, C. M.; Pereira, M. M.; Teixeira, M.; Costa, T.; Jones, G. H.; Henriques, A. O. Molecular and biochemical characterization of a highly stable bacterial laccase that occurs as a structural component of the *Bacillus subtilis* endospore coat. *J. Biol. Chem.* **2002**, *277*(21), 18849–18859.

30. Li, K.; Xu, F.; Eriksson, K. E. L. Comparison of fungal laccases and redox mediators in oxidation of a nonphenolic lignin model compound. *Appl. Environ. Microbiol.* **1999**, *65*, 2654–2660.

31. Bourbonnais, R.; Paice, M. G.; Freiermuth, B.; Bodie, E.; Borneman, S. Reactivities of various mediators and laccases with kraft pulp and lignin model compounds. *Appl. Environ. Microbiol.* **1997**, *63*, 4627–4632.

32. Srebotnik, E.; Hammel, K. E. Degradation of nonphenolic lignin by the laccase/1-hydroxybenzotriazole system. *J. Biotechnol.* **2000**, *81*(2–3), 179–188.

33. Bourbonnais, R.; Paice, M. G. Enzymatic delignification of kraft pulp using laccase and a mediator. *Tappi J.* **1996**, *79*, 199–204.

34. Balakshin, M.; Chen, C.; Gratzl, J. S.; Kirkman, A. G.; Jakob, H.; Chen, C. L. Biobleaching of pulp with dioxygen in the laccase-mediator system. Part 1. Kinetics of delignification. *Holzforschung* **1999**, *54*(4), 390–396.

35. Muheim, A.; Fiechter, A.; Harvey, P. J.; Schoemaker, H. E. On the mechanism of oxidation of non-phenolic lignin model compounds by the laccase-abts couple. *Holzforschung* **1992**, *46*(2), 121–126.

36. Johannnes, C.; Majcherczyk, A. Natural mediators in the oxidation of polycyclic aromatic hydrocarbons by laccase mediator systems. *Appl. Environ. Microbiol.* **2000**, *66*, 524–528.

37. Call, H. P.; Mucke, I. History, overview and applications of mediated lignolytic systems, especially laccase-mediator-systems (Lignozym® process). *J. Biotechnol.* **1997**, 53(2–3), 163–202.

38. Li, K.; Prabhu, G. N.; Cooper, D. A.; Xu, F.; Elder, T.; Eriksson, K.-E. L. Development of new laccase-mediators for pulp bleaching. In: Argyropoulos, D. S. (Ed.) *Oxidative Delignification Chemistry*. American Chemical Society, Washington, **2001**, 400–412.

39. Call, H. P.; Mucke, I. Enzymatic bleaching of pulps with the laccase-mediator-system (lms). In: *Advances in Pulp and Papermaking* (ed.); American Institute of Chemical Engineers, New York, Symposium series #307, **1995**; 91, 38–52.

40. Bourbonnais, R.; Paice, M. G. Demethylation and delignification of kraft pulp by *Trametes versicolor* laccase in the presence of 2,2′-azino*bis*-(3-ethylbenzthiazoline-6-sulphonate). *Appl. Microbiol. Biotechnol.* **1992**, *36*(6), 823–827.

41. Shleev, S. V.; Khan, I. G.; Gazaryan, I. G.; Morozova, O. V.; Yaropolov, A. I. Novel laccase redox mediators. *Appl. Biochem. Biotechnol.* **2003**, *111*, 167–183.

42. Paice, M. G.; Bourbonnais, R.; Reid, I. D.; Archibald, F.; Jurasek, L. Oxidative bleaching enzymes: A review. *J. Pulp Paper Sci.* **1995**, *21*(8), J280–J284.

43. Baiocco, P.; Barreca, A. M.; Fabbrini, M.; Galli, C.; Gentili, P. Promoting laccase activity towards non-phenolic substrates: A mechanistic investigation with some laccase-mediator systems. *Org. Biomol. Chem.* **2003**, *1*, 191–197.

44. d'Acunzo, F.; Galli, C.; Masci, B. Oxidation of phenols by laccase and laccase-mediator systems. *Eur. J. Biochem.* **2002**, *269*, 5330–5335.

45. Pilz, R.; Hammer, E.; Schauer, F.; Kragl, U. Laccase-catalysed synthesis of coupling products of phenolic substrates in different reactors. *Appl. Microbiol. Biotechnol.* **2003**, *60*, 708–712.

46. Knutson, K.; Ragauskas, A. Laccase-mediator biobleaching applied to a direct yellow dyed paper. *Biotechnol. Prog.* **2004**, *20*, 1893–1896.

47. Bourbonnais, R.; Rochefort, D.; Paice, M. G.; Renaud, S.; Leech, D. Transition metal complexes: A new class of laccase mediators for pulp bleaching. *Tappi J.* **2000** (October), 68.

48. Eggert, C.; Temp, U.; Dean, J. F. D.; Eriksson, K. E. L. A fungal metabolite mediates degradation of non-phenolic lignin structures and synthetic lignin by laccase. *FEBS Lett.* **1996**, *391*(1–2), 144–148.

49. Wong, K. K. Y.; Anderson, K. B.; Kibblewhite, R. P. Effects of the laccase-mediator system on the handsheet properties of two high kappa kraft pulps. *Enzyme Microb. Technol.* **1999**, *25*, 125–131.

50. Gronqvist, S.; Buchert, J.; Rantanen, K.; Viikari, L.; Suurnakki, A. Activity of laccase on unbleached and bleached thermomechanical pulp. *Enzyme Microb. Technol.* **2003**, *32*, 439–445.

51. Sealey, J.; Ragauskas, A. J. Residual lignin studies of laccase-delignified kraft pulps. *Enzyme Microb. Technol.* **1998**, *23*(7–8), 422–426.

52. Sealey, J.; Ragauskas, A. J.; Elder, T. J. Investigations into laccase-mediator delignification of kraft pulps. *Holzforschung* **1999**, *53*(5), 498–502.

53. Camarero, S.; Garcia, O.; Vidal, T.; Colom, J.; del Rio, J. C.; Gutierrez, A.; Gras, J. M.; Monje, R.; Martinez, M. J.; Martinez, A. T. Efficient bleaching of non-wood high-quality paper pulp using laccase-mediator system. *Enzyme Microb. Technol.* **2004**, *35*, 113–120.

54. Huttermann, A.; Mai, C.; Kharazipour, A. Modification of lignin for the production of new compounded materials. *Appl. Microbiol. Biotechnol.* **2001**, *55*, 387–394.

55. Wong, K. K. Y.; Richardson, J. D.; Mansfield, S. D. Enzymatic treatment of mechanical pulp fibers for improving papermaking properties. *Biotechnol. Prog.* **2000**, *16*(6), 1025–1029.

56. Lund, M.; Felby, C. Wet strength improvement of unbleached kraft pulp through laccase catalyzed oxidation. *Enzyme Microb. Technol.* **2001**, *28*, 760–765.

57. Felby, C.; Pedersen, L. S.; Nielsen, B. R. Enhanced auto adhesion of wood fibers using phenol oxidases. *Holzforschung* **1997**, *51*(3), 281–286.

58. Mansfield, S. D. Laccase impregnation during mechanical pulp processing-improved refining efficiency and sheet strength. *Appita J.* **2002**, *55*, 49–53.

59. Felby, C.; Hassingboe, J.; Lund, M. Pilot-scale production of fiberboards made by laccase oxidized wood fibers: Board properties and evidence for cross-linking of lignin. *Enzyme Microb. Technol.* **2002**, *31*, 736–741.

60. Felby, C.; Olesen, P. O.; Hansen, T. T. Laccase catalyzed bonding of wood fibers. In: Eriksson, K. E.; Cavaco-Paulo, A. (Eds.) *Enzyme Applications in Fiber Processing.* American Chemical Society, Washington, **1998**, Symposium Series 687, 88–98.

61. Mai, C.; Schormann, W.; Huettermann, A. The effect of ions on the enzymatically induced synthesis of lignin graft copolymers. *Enzyme Microb. Technol.* **2001**, *28*(4–5), 460–466.

62. Huttermann, A.; Majcherczyk, A.; Braun, L. A.; Mai, C.; Fastenrath, M.; Kharazipour, A.; Huttermann, J.; Huttermann, A. H. Enzymatic activation of lignin leads to an unexpected copolymerization with carbohydrates. *Naturwissenschaften* **2000**, *87*, 539–541.

63. Mai, C.; Majcherczyk, A.; Huttermann, A. Chemo-enzymatic synthesis and characterization of graft copolymers from lignin and acrylic compounds. *Enzyme Microb. Technol.* **2000**, *27*(1–2), 167–175.

64. Mai, C.; Schormann, W.; Huttermann, A. Chemo-enzymatically induced copolymerization of phenolics with acrylate compounds. *Appl. Microbiol. Biotechnol.* **2001**, *55*(2), 177–186.

65. Mai, C.; Milstein, O.; Huttermann, A. Fungal laccase grafts acrylamide onto lignin in presence of peroxides. *Appl. Microbiol. Biotechnol.* **1999**, *51*, 527–531.

66. Mai, C.; Milstein, O.; Hutterman, A. Chemoenzymatical grafting of acrylamide onto lignin. *J. Biotechnol.* **2000**, *79*, 173–183.

67. Kenealy, W.; Klungness, J.; Tshabalala, M. A.; Horn, E.; Akhtar, M.; Gleisner, R.; Buschle-Diller, G. Modification of lignocellulosic materials by laccase. TAPPI Fall Technical Conference, Chicago, IL, Oct 26–30, **2003**; Abstract 3381.

68. Chandra, R. P.; Lehtonen, L. K.; Ragauskas, A. J. Modification of high lignin content kraft pulps with laccase to improve paper strength properties. 1. Laccase treatment in the presence of gallic acid. *Biotechnol. Prog.* **2004**, *20*, 255–261.

69. Chandra, R. P.; Ragauskas, A. J. Evaluating laccase-facilitated coupling of phenolic acids to high-yield kraft pulps. *Enzyme Microb. Technol.* **2002**, *30*(7), 855–861.

70. Chandra, R. P.; Ragauskas, A. J. Laccase: The renegade of fiber modification. Tappi Pulping Conference, Tappi press, Atlanta, GA, **2001**.

71. Mueller, M.; Shi, C. Laccase for denim processing. *AATCC Rev.* **2001**, *1*(7), 4–5.

72. Tzanov, T.; Basto, C.; Guebitz, G. M.; Cavaco-Paulo, A. Laccases to improve the whiteness in a conventional bleaching of cotton. *Macromol. Mater. Eng.* **2003**, *288*, 807–810.

73. Ossala, M.; Galante, Y. M. Scouring of flax rove with the aid of enzymes. *Enzyme Microb. Technol.* **2004**, *34*, 177–186.

74. Shin, H.; Cavaco-Paulo, A., In-situ enzymatically prepared polymers for wool coloration. *Macromol. Mater. Eng.* **2001**, *286*, 691–694.

75. Aaslyng, D., Sorensen, N. H., Rorbaek, K. Laccases with improved dyeing properties. US 5948121, **1998**.

76. Paice, M. G.; Bourbonnais, R. Non-chlorine bleaching of kraft pulp. US Patent 691193, **1997**.

77. Paice, M. G.; Bourbonnais, R.; Reid, I. D. Bleaching kraft pulps with oxidative enzymes and alkaline hydrogen peroxide. *Tappi J.* **1995**, *78*(9), 161–169.

78. Sigoillot, J. C.; Petit, C. M.; Herpoel, I.; Joseleau, J. P.; Ruel, K.; Kurek, B.; de Choudens C.; Asther, M. Energy saving with fungal enzymatic treatment of industrial poplar alkaline peroxide pulps. *Enzyme Microb. Technol.* **2001**, *29*(2–3), 160–165.

79. Buschle-Diller, G. Oxidoreductases for textile applications. INTB Intern. Network Text. Biotechnol., Graz, Austria, Jun 13–16, **2004**.

80. Mansfield, S. D.; De, J. E.; Saddler, J. N. Cellobiose dehydrogenase, an active agent in cellulose depolymerization. *Appl. Environ. Microbiol.* **1997**, *63*(10), 3804–3809.

81. Henriksson, G.; Johansson, G.; Pettersson, G. A critical review of cellobiose dehydrogenases. *J. Biotechnol.* **2000**, *78*, 93–113.

82. Pramod, K. Liquid laundry detergents containing stabilized glucose–glucose oxidase system for hydrogen peroxide generation. US 5288746, **1994**.

83. Thoen, C. A. J.; Fredj, A.; Labeque, R. Detergent compositions inhibiting dye transfer in washing. EP 596186, **1994**.

84. Buschle-Diller, G.; Yang, X. D.; Yamamoto, R. Enzymatic bleaching of cotton fabric with glucose oxidase. *Text. Res. J.* **2001**, *71*(5), 388–394.

85. Tzanov, T.; Costa, S. A.; Guebitz, G. M.; Cavaco-Paulo, A. Hydrogen peroxide generation with immobilized glucose oxidase for textile bleaching. *J. Biotechnol.* **2002**, *93*, 87–94.

86. Clarke, J. H.; Rixon, J. E.; Ciruela, A.; Gilbert, H. J.; Hazlewood, G. P. Family-10 and family-11 xylanases differ in their capacity to enhance the bleachability of hardwood and softwood paper pulps. *Appl. Microbiol. Biotechnol.* **1997**, *48*(2), 177–183.

87. Viikari, L.; Ranva, M.; Kantelinen, A.; Sanquist, J.; Linko, M. Bleaching with enzymes. Third International Conference in Biotechnology in the Pulp and Paper Industry, Stockholm, Sweden, **1986**, 67–69.

88. Rizzatti, A. C. S.; Sandrim, V. C.; Jorge, J. A.; Terenzi, H. F.; de Lourdes, M.; Polizeli, M. L. T. M. Influence of temperature on the properties of the xylanolytic enzymes of the thermotolerant fungus *Aspergillus phoenicis*. *J. Ind. Microbiol. Biotechnol.* **2004**, *31*, 88–93.

89. Bocchini, D. A.; Damiano, V. B.; Gomes, E.; Da, S. R. Effect of *Bacillus circulans* d1 thermostable xylanase on biobleaching of eucalyptus kraft pulp. *Appl. Biochem. Biotechnol.* **2003**, *106*, 393–401.

90. Kumar, B. K.; Balakrishnan, H.; Rele, M. V. Compatibility of alkaline xylanases from and alkaliphilic *Bacillus* ncl (87-6-10) with commercial detergents and proteases. *J. Ind. Microbiol. Biotechnol.* **2004**, *31*, 83–87.

91. Wong, K. K. Y.; de Jong, E.; Saddler, J. N.; Allison, R. W. Mechanisms of xylanase aided bleaching of kraft pulp. *Appita J.* **1997**, *50*(6), 509–518.

92. Kantelinen, A.; Hortling, B.; Sundquist, J.; Linko, M.; Viikari, L. Proposed mechanism of the enzymatic bleaching of kraft pulp with xylanases. *Holzforschung* **1993**, *47*, 318–324.

93. Hortling, B.; Korhonen, M.; Buchert, J.; Sundquist, J.; Viikari, L. The leachability of lignin from kraft pulps after xylanase treatment. *Holzforschung* **1994**, *48*, 441–446.

94. de Jong, E.; Wong, K. K. Y.; Saddler, J. N. The mechanism of xylanase prebleaching of kraft pulp: An examination of using model pulps prepared by depositing lignin and xylan on cellulose fibers. *Holzforschung* **1997**, *51*, 19–26.

95. Chakar, F. S.; Allison, L.; Ragauskas, A.; McDonough, T. J.; Sezgi, U. Influence of hexenuronic acids on U.S. bleaching operations. *Tappi J.* **2000** *83*(November), 62.

96. Dence, C. W.; Reeve, D. W. In: *Pulp Bleaching—Principles and Practice*. Tappi Press, Atlanta, **1996**, 365–377.

97. Viikari, L.; Kantelinen, A.; Sundquist, J.; Linko, M. Xylanases in bleaching: from an idea to the industry. *FEMS Microbiol. Rev.* **1994**, *13*, 335–350.

98. Jiang, Z. H.; Lierop, B. V.; Berry, R. Hexenuronic acid groups in pulping and bleaching chemistry. *Tappi J.* **2000**, *83*(1), 167–175.

99. Elegir, G.; Sykes, M.; Jeffries, T. W. Differential and synergistic action of *Streptomyces* endoxylanases in prebleaching of kraft pulps. *Enzyme Microb. Technol.* **1995**, *17*(10), 954–959.

100. Bermek, H.; Li, K.; Eriksson, K. E. L. Pulp bleaching with manganese peroxidase and xylanase: A synergistic effect. *Tappi J.* **2000**, *83*(10), 69.

101. Beg, Q. K.; Kapoor, M.; Mahajan, L.; Hoondal, G. S. Microbial xylanases and their indus-
     trial applications: A review. *Appl. Microbiol. Biotechnol.* **2001**, *56*(3–4), 326–338.
102. Bajpai, P.; Bajpai, P. K. Deinking with enzymes: A review. *Tappi J.* **1998**, *81*(12),
     111–117.
103. Clarke, J. H.; Davidson, K.; Rixon, J. E.; Halstead, J. R.; Fransen, M. P.; Gilbert, H. J.;
     Hazlewood, G. P. A comparison of enzyme-aided bleaching of softwood paper pulp using
     combinations of xylanase, mannanase and α-galactosidase. *Appl. Microbiol. Biotechnol.*
     2000, *53*(6), 661–667.
104. Buchert, J.; Salminen, J.; Siika, A. M.; Ranua, M.; Viikari, L. The role of *Trichoderma
     reesei* xylanase and mannanase in the treatment of softwood kraft pulp prior to bleaching.
     Holzforschung **1993**, *47*, 473–478.
105. Suurnakki, A.; Clark, T.; Allison, D.; Viikari, L.; Buchert, J. Xylanase and mannanase-aided
     ecf and tcf bleaching. *Tappi J.* **1996**, *79*(7), 111–117.
106. Suurnakki, A.; Heijnesson, A.; Buchert, J.; Tenkanen, M.; Viikari, L.; Westermark, U.
     Location of xylanase and mannanase action in kraft fibers. *J. Pulp Paper Sci.* **1996**, *22*(3),
     J78–J83.
107. Saake, B.; Clark, T.; Puls, J. Investigations on the reaction mechanism of xylanases and
     mannanases on sprucewood chemical pulps. *Holzforschung* **1995**, *49*, 60–68.
108. Sunna, A.; Gibbs, M. D.; Chin, C. W.; Nelson, P. J.; Bergquist, P. L. A gene encoding a
     novel multidomain β-1,4-mannanase from *Caldibacillus cellulovorans* and action of the
     recombinant enzyme on kraft pulp. *Appl. Environ. Microbiol.* **2000**, *66*(2), 664–670.
109. Kansoh, A. L.; Nagieb, Z. A. Xylanase and mannanase enzymes from *Streptomyces galbus*
     nr and their use in biobleaching of softwood kraft pulp. *Antonie van Leeuwenhoek* **2004**,
     85, 103–114.
110. Gutierrez, A.; del Rio, J. C.; Martinez, M. J.; Martinez, A. T. The biotechnological control
     of pitch in paper pulp manufacturing. *Trends Biotechnol.* **2001**, *19*(9), 340–348.
111. Pere, J.; Siika, A. M.; Buchert, J.; Viikari, L. Effects of purified *Trichoderma reesei* cellu-
     lases on the fiber properties of kraft pulp. *Tappi J.* **1995**, *78*(6), 71–78.
112. Pommier, J. C.; Fuentes, J. L.; Goma, G. Using enzymes to improve the process and the
     product quality in the recycled paper industry. *Tappi J.* **1989**, *72*(6), 187–191.
113. Jackson, L. S.; Heitmann, J. A.; Joyce, T. W. Enzymatic modifications of secondary fiber.
     Tappi J. **1993**, *76*(3), 147–154.
114. Heise, O. U.; Fineran, W. G.; Unwin, J. P.; Sykes, M. S.; Klungness, J. H.; Abubakr,
     S. Industrial scale-up of enzyme enhanced deinking of non-impact printed toners. Tappi
     Pulping Conference, Tappi Press, Altanta, **1995**, 349–354.
115. Jeffries, T. W.; Klungness, J. H.; Sykes, M. S.; Rutledge-Cropsey, K. Comparison of enzyme-
     enhanced with conventional deinking of xerographic and laser-printed paper. *Tappi J.* **1994**
     77(April), 173–179.
116. Klungness, J. H.; Sykes, M. S.; Jeffries, T. W.; Abubakr, S. Enzyme enhanced deinking
     of toners. In: Doshi, D. (Ed.) *Paper Recycling Challenge Vol II—Deinking.* Doshi and
     Associates, Appleton, WI, **1997**, 155–160.
117. Rutledge-Cropsey, K.; Klungness, J. H.; Abubakr, S. Performance of enzymatically deinked
     wastepaper on papermachine runnability. Tappi Pulping Conference, Tappi Press, Atlanta,
     1995, 639–643.
118. Welt, T.; Dinus, R. J. Enzymatic deinking: Effectiveness and mechanisms. *Wochenb. Pa-
     pierfabrik.* **1998**, *126*(9), 396–407.
119. Welt, T.; Dinus, R. J. Enzymatic deinking—A review. *Prog. Paper Recycling* **1995** 4(Febru-
     ary), 36–45.

120. Moran, B. R. Enzyme treatment improves refining efficiency, recycled fiber freeness. *Pulp Paper* **1996** *70*(September), 119–121.
121. Zeyer, C.; Joyce, T. W.; Rucker, J. W.; Heitmann, J. A. Enzymatic deinking of cellulose fabric: A model study for enzymatic paper deinking. *Prog. Paper Recycling* **1993** *3*(November), 36–44.
122. Sykes, M. S.; Klungness, J. H.; Abubakr, S.; Tan, F. Upgrading recovered paper with enzyme pre-treatment and pressurized peroxide bleaching. *Prog. Paper Recycling* **1996**, *5*(August), 39–45.
123. Elegir, G.; Panizza, E.; Canetti, M. Neutral-enzyme-assisted deinking of xerographic office waste with a cellulase–amylase mixture. *Tappi J.* **2000**, *83*(11), 71.
124. Park, J. W.; Park, K. N. Biological de-inking of waste paper using modified cellulase with polyoxyethylene. *Biotechnol. Tech.* **1999**, *13*, 49–53.
125. Morkbak, A. L.; Zimmermann, W. Deinking of mixed office paper, old newspaper and vegetable oil-based ink printed paper using cellulases, xylanases and lipases. *Prog. Paper Recycling* **1998** *7*(February), 14–21.
126. Sanciolo, P.; Warnock, H.; Harding, I.; Forbes, L.; Lonergan, G. Microscopy study: Liberation of ink from fibers during enzymatic deinking of mixed office papers. *Prog. Paper Recycling* **2000** *9*(May), 22–30.
127. Sykes, M. S.; Klungness, J. H.; Geisner, R.; Abubakr, S. Stickie removal using neutral enzymatic pulping and pressure screening. Tappi Recycling Symposium, Tappi Press, Atlanta, **1998**, 291–296.
128. Schulein, M.; Outtrup, H.; Jorgensen, P. L.; Bjornvad, M. E. Alkaline xyloglucanases from *Bacillus* suitable for fabric detergents. US 9902663, **1999**.
129. Nickel, D.; Bianconi, P.; Voelkel, T.; Speckmann, H.-D.; Jekel, M. Textile care agent containing cellulase and a color-fixing agent. IntPA 2001074982, **2001**.
130. Galante, Y. M.; Formantici, C. Enzyme applications in detergency and in manufacturing industries. *Curr. Org. Chem.* **2003**, *7*(13), 1399–1422.
131. Levene, R.; Shakkour, G. Wool fibers of enhanced luster obtained by enzymic descaling. *J. Soc. Dyers Colour* **1995**, *111*(11), 352–359.
132. Guo, W.; Zhang, J.; Cheng, L. Silk cold padding and whitening method. *Faming Zhuanli Shenqing Gongkai Shuomingshu* **2001**, CNXXEV CN1284574 A 20010221.
133. Pedersen, G. L.; Screws, G. A.; Cedroni, D. M. Biopolishing of cellulosic textiles. *Tinctoria* **1993**, *90*(8), 59–63.
134. Cortez, J. M.; Ellis, J.; Bishop, D. P. Cellulase finishing of woven, cotton fabrics in jet and winch machines. *J. Biotechnol.* **2001**, *89*(2, 3), 239–245.
135. Ramkumar, S. S.; Abdalah, G. Surface characterization of enzyme treated fabrics. *Colourage* **2001**, *48*(4), 15–16, 24.
136. Auterinen, A.; Carreras, D.; Navarro, M. Developments in enzyme technology for textile application. *Rev. Quim. Text.* **2000**, *148*(34), 36–37, 40–42, 44.
137. Heikinheimo, L.; Cavaco-Paulo, A.; Nousiainen, P.; Siika-aho, M.; Buchert, J. Treatment of cotton fabrics with purified *Trichoderma reesei* cellulases. *J. Soc. Dyers Colour.* **1998**, *114*(7/8), 216–220.
138. Cavaco-Paulo, A.; Guebitz, G. M. (Eds.) *Textile Processing with Enzymes*. Woodhead Publ., Cambridge, U.K., **2000**, 86–119.
139. Taylor, J.; Fairbrother, A., Tencel—it's more than just peachskin. *J. Soc.Dyers Colour.* **2000**, *116*(12), 381–384.
140. Gandhi, K.; Burkinshaw, S. M.; Taylor, J. M.; Collins, G. W. A novel route for obtaining a "peach skin effect" on lyocell and its blends. *AATCC Rev.* **2002**, *2*(4), 48–52.

141. Ibrahim, N. A.; El-Hossamy, M.; Morsy, M. S.; Eid, B. M. Development of new eco-friendly options for cotton wet processing. *J. Appl. Polym. Sci.* **2004**, *93*(4), 1825–1836.

142. Losonczi, A.; Csiszar, E.; Szakacs, G.; Kaarela, O. Bleachability and dyeing properties of biopretreated and conventionally scoured cotton fabrics. *Text. Res. J.* **2004**, *74*(6), 501–508.

143. Calafell, M.; Garriga, P. Effect of some process parameters in the enzymatic scouring of cotton using an acid pectinase. *Enzyme Microb. Technol.* **2004**, *34*(3–4), 326–331.

144. Traore, M. K.; Buschle-Diller, G. Environmentally friendly scouring processes. *Text. Chem. Color. Am. Dyestuff Rep.* **2000**, *32*(12), 40–43.

145. Waddell, R. B. Bioscouring of cotton: Commercial applications of alkaline stable pectinase. *AATCC Rev.* **2002**, *2*(4), 28–30.

146. Takagishi, T.; Yamamoto, R.; Kikuyama, K.; Arakawa, H. Design and application of continuous bio-scouring machine. *AATCC Rev.* **2001**, *1*(8), 32–34.

147. Terpe, K. Overview of tag protein fusions: From molecular and biochemical fundamentals to commercial systems. *Appl. Microbiol. Biotechnol.* **2003**, *60*(5), 523–533.

148. Shoseyov, O.; Warren, R. A. J. Cellulose binding domains—a novel fusion technology for efficient, low cost purification and immobilization of recombinant proteins. *Innovations* **1997**, *7*, 1–3.

149. Cavaco, P. A.; Morgado, J.; Andreaus, J.; Kilburn, D. Interactions of cotton with cbd peptides. *Enzyme Microb. Technol.* **1999**, *25*(8–9), 639–643.

150. Xiao, Z.; Gao, P.; Qu, Y.; Wang, T. Cellulose-binding domain of endoglucanase III from *Trichoderma reesei* disrupting the structure of cellulose. *Biotechnol. Lett.* **2001**, *23*(9), 711–715.

151. Pala, H.; Lemos, M. A.; Mota, M.; Gama, F. M. Enzymatic upgrade of old paperboard containers. *Enzyme Microb. Technol.* **2001**, *29*, 274–279.

152. Nigmatullin, R.; Lovitt, R.; Wright, C.; Linder, M.; Nakari-Setala, T.; Gama, M. Atomic force microscopy study of cellulose surface interaction controlled by cellulose binding domains. *Colloids Surfaces B: Biointerfaces* **2004**, *35*, 125–135.

153. Boraston, A. B.; McLean, B. W.; Chen, G.; Li, A.; Warren, R. A. J.; Kilburn, D. G. Co-operative binding of triplicate carbohydrate-binding modules from a thermophilic xylanase. *Mol. Microbiol.* **2002**, *43*(1), 187–194.

154. Xu, Z.; Bae, W.; Mulchandani, A.; Mehra, R. K.; Chen, W. Heavy metal removal by novel cbd-ec20 sorbents immobilized on cellulose. *Biomacromolecules* **2002**, *3*(3), 462–465.

155. Degani, O.; Gepstein, S.; Dosoretz, C. G. A new method for measuring scouring efficiency of natural fibers based on the cellulose-binding domain-β-glucuronidase fused protein. *J. Biotechnol.* **2004**, *107*, 265–273.

156. Brumer III, H.; Zhou, Q.; Baumann, M. J.; Carlsson, K.; Teeri, T. Activation of crystaline cellulose surfaces through chemoenzymatic modification of xyloglucan. *J. Am. Chem. Soc.* **2004**, *126*, 5715–5721.

157. Lima, D. U.; Oliveira, R. C.; Buckeridge, M. S. Seed storage hemicelluloses as wet end additives in papermaking. *Carbohydr. Polym.* **2003**, *52*, 367–373.

Chapter 13

THE ATTRACTION OF MAGNETICALLY SUSCEPTIBLE PAPER

Douglas G. Mancosky and Lucian A. Lucia

Institute of Paper Science and Technology, Georgia Institute of Technology, 500 10th St. NW, Atlanta, Georgia 30332-0620, U.S.A.

Abstract

We have imparted magnetic susceptibility to lignocellulosic fibers by adding iron powder to the fibers during hydrogen peroxide bleaching chemistry. We have, therefore, generated carboxylic acid groups in the fibers by deliberately inducing cellulose degradation through Fenton catalysis of the hydrogen peroxide during the chemical oxidation process at a specified level of iron. The iron particles consequently have an exposed layer of iron oxide that allows ionic neutralization of the negatively charged fiber acid groups. After removal of non-attached excess iron, these fibers have been cast into two-dimensional sheets with two different original iron concentrations and tested for physical and chemical properties. Physical tests included tensile, zero-span tensile, caliper, and surface resistivity. Chemical tests included surface charge, kappa, and viscosity. Scanning electron microscopy (SEM) and inductively coupled plasma (ICP) emission spectroscopy were also conducted. Remarkably, the magnetically susceptible sheets with incorporated iron were able to retain a tensile strength similar to the unbleached sheets despite attenuation in fiber strength. This is likely due to a chemical refining phenomenon that allowed for increased fiber–fiber bonding. The introduction of the retained iron also significantly alters the surface resistivity of the paper sheets. Such fibers may have a use in applications where charge conduction or dispersion is necessary.

Keywords: magnetic susceptibility; lignocellulosics; iron

13.1 Introduction

Imparting specific functionality to lignocellulosic fibers potentially offers great promise to increasing the attractiveness and utility of related materials. Recently, a number of research groups have focused on chemically manipulating these fibers for controlled applications or unique functionality. For

209

J. V. Edwards et al. (eds.), Modified Fibers with Medical and Specialty Applications, 209–214.

example, lignocellulosics, in general, have been at the "core" of several material processing applications such as lignosulfate-based hydrogels, polyurethane foams, bimorphous ceramics, photocatalysis supports, and thermoplastic extrusions.

On a similar front, fiber modification with a view toward developing new functionality is currently witnessing a renaissance. Choplin has demonstrated the feasibility of using cellulose as a support for a catalyst (Pd-based) for allylic alkylations [1]. Basso has shown lignocellulosics as feasible biosorbents of trace toxic metals [2]. Ghosh was able to photofunctionalize fibers with photoactive acids for grafting and property modification [3]. Induction of magnetic susceptibility in lignocellulosics, however, has received very little research attention.

The work presented here has examined the nature of lignocellulosics as platforms that support the force of magnetism. Recently, we have explored the induction of magnetic susceptibility in lignocellulosic fibers through the generation of fiber carboxylic groups as a result of iron-catalyzed hydrogen peroxide bleaching. Using titration measurements as an elementary probe for the accretion of carboxylic acid functionalities, we observed an increase of ca. 20% in acid functionalities on the surface (55 µeq/g of pulp). The increase is lower than expected due to a substantial loss in carbohydrates through the degradation process. We have found from previous work that hydrogen peroxide bleaching is typically a very effective method to introduce carboxylic acid groups in kraft pulps.

13.2 Experimental

All work was conducted using standard hydrogen peroxide bag bleaches. Bleaches were conducted at 15% peroxide for 2 h in which trace peroxide residuals were determined after the bleach. All bag bleaches were outfitted with several bags due to extensive gas production during the bleach. All runs were done at temperatures of $90\,^{\circ}C$ and conducted at 10% iron. A caustic solution of pH 12 was used for all make-up water, which was added to maintain all pulp-water solutions at a 10% solid level (mass of pulp/mass of pulp and water). A base-bleached sample was designated, as the H_2O_2 sample had the iron separated by several days of mixing on a magnetic stir plate in which the iron was extracted. Iron could not be easily extracted with normal washing. "25%" iron level sheets (m/m) were made by adding additional iron powder to the bleached pulp to achieve the mass balance ratio required. Tensile strength was tested using the TAPPI standard method on random handsheets. SEM was conducted using standard operational conditions. Surface carboxylic acid group levels were determined by standard potentiometric acid titration procedures. All percentages of iron were determined *via* ICP. Volume and surface resistivity

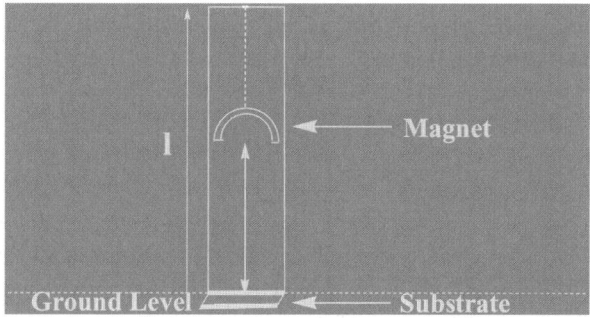

Figure 13.1. Setup for measure of magnetic susceptibility

were determined by ASTM D4949-94 and determination of DC resistivity of writing paper (Keithley Method). Magnetic susceptibility was determined using the setup depicted in Figure 13.1. The interpolar distance, l, for threshold of attractive adherence to the poles of the magnet was measured.

13.3 Results

Once it was discovered that the iron was intractably bound to the sheets and could not be readily removed, a study was designed to determine the properties of sheets made from the magnetically susceptible fibers. Initial efforts focused on studying the nature of the adherence of the iron particles to the fibers using SEM. The iron particle distributions appeared heterogeneous and also were found to bound to fiber surface as opposed to physically entrapped between fibers. A sample micrograph can be seen in Figure 13.2. The iron particles were found to be physico-chemically adsorbed to the surface of the fibers. These fibers were washed extensively with deionized water, but demonstrated fair adsorption of the small iron particles across the fiber surface.

The next step in our investigations was to determine the levels of iron contained in each pulp. All iron addition levels were made on the basis of unbleached fiber. It was clear there was significant degradation and yield loss during bleaching since the final levels of iron, as determined by ICP, were significantly higher than the projected amounts based on raw fiber. It was also clear that our method of removing the iron from the bleached pulp over several days of magnetic stirring was effective since the levels of iron contained could be reduced to the levels found in the base pulp. The ICP measurements are shown in Table 13.1.

The next step was to determine the impact of degradation and iron incorporation in the sheets on the sheet strength. It was determined that the overall sheet strength was not significantly affected. This was most likely due to a chemical

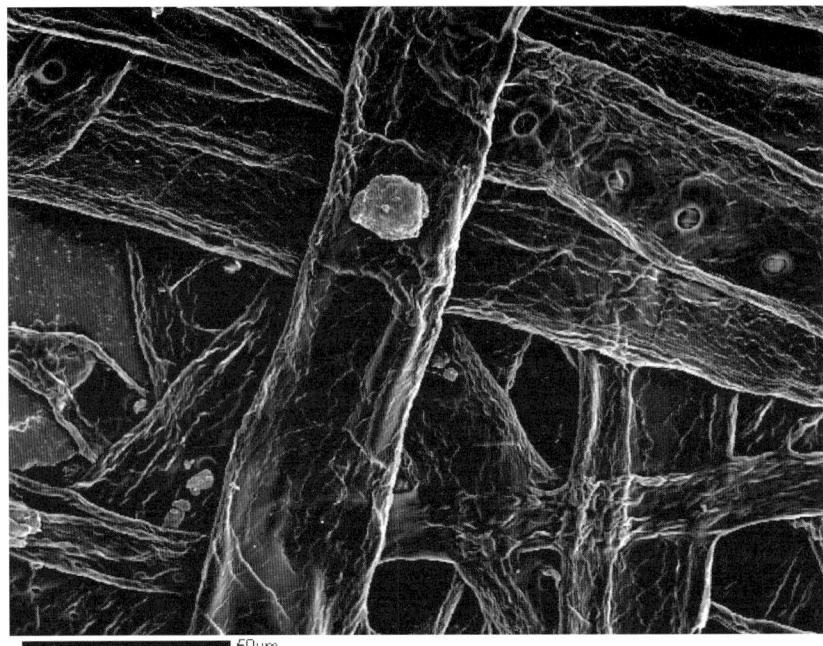

20% iron powder, shiny side down, 600X

Figure 13.2. A typical SEM micrograph of a 20% iron powder sheet at 600× magnification

refining effect allowing for greater fiber bonding, even though individual fiber strength was most likely decreased. Tensile measurements can be seen in Figure 13.3.

The finding in Figure 13.4 was remarkable and extremely important for the further evaluation of this process. Although not unprecedented, it demonstrated that bonding could be improved *via* chemical means.

It was important to understand the level of magnetic susceptibility imparted to the pulp and compare this to a full-metal substrate. As expected, when the percentage of iron was increased, the magnetic susceptibility was increased.

Table 13.1. The iron addition levels that were found in the pulps from ICP evaluation

Substrate	Fe (mg/kg)	Percentage (iron/pulp)
Base pulp	3830	0.4
H_2O_2 bleached	2160	0.2
10% Fe^0 H_2O_2 bleached	130,000	13
25% Fe^0 H_2O_2 bleached	317,000	32

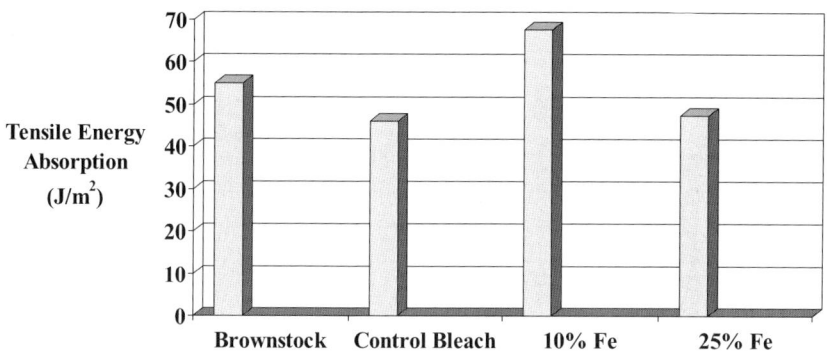

Figure 13.3. Tensile energy absorption comparison

Interestingly, the magnetic susceptibility was proportional to the level of iron. This work can be seen in Figure 13.4.

In addition to magnetic susceptibility, the iron addition also imparted another interesting property: a significant decrease in the electrical resistivity. The volume resisitivity was cut in half for the 10% sample and almost to one-third in the 25% sample. The surface resistance was even more greatly impacted. The bleached sample with the iron removed actually had a higher resistance than the base stock. The 10% and 25% iron samples had near equal resistivity that was about 1/250th of the original base stock. This could have significant implications on new products where electrical properties are important such as computers. A graph comparing surface and volume resistivity for all samples can be seen in Figure 13.5.

From these results, it is possible to impart meaningful magnetic susceptibility to pulp using hydrogen peroxide bleaching, without significant changes in tensile strength. This iron addition has the welcome side effect of changing

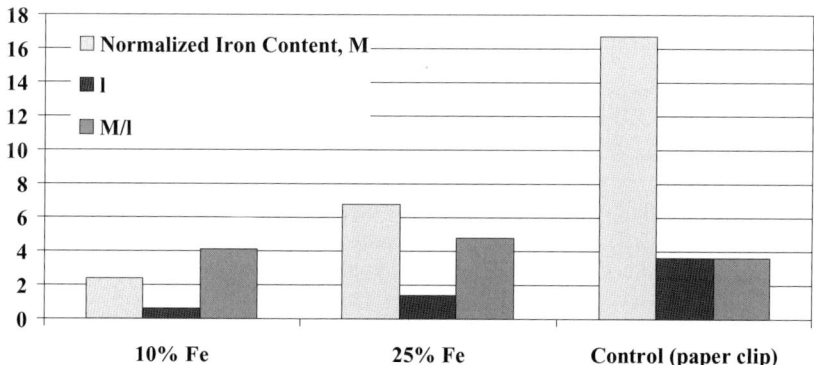

Figure 13.4. Comparison of magnetic susceptibility

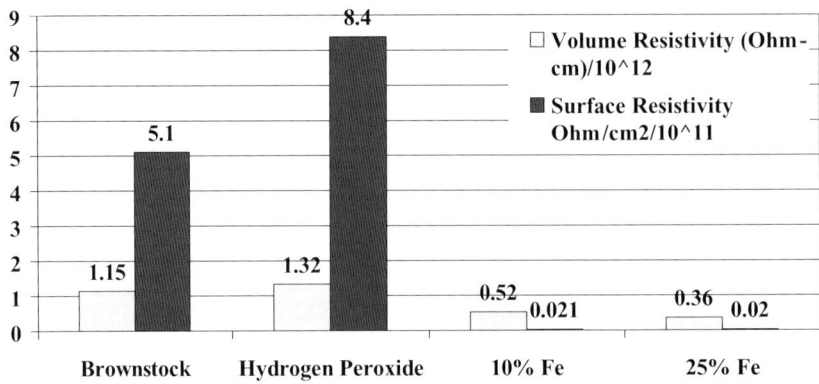

Figure 13.5. Comparison of surface and volume resistivity

the electrical resistance of the sheet. Such fiber modification could allow for use in unique end products and increase pulp value.

13.4 Conclusions

Iron powder was successfully incorporated into peroxide-bleached sheets resulting in magnetic susceptibility. The level of magnetic susceptibility was proportional to the level of iron incorporated. The magnetically susceptible sheets, with incorporated iron powder, were able to retain a tensile strength similar to the base (unbleached) stock. The magnetically susceptible sheets also had dramatically decreased volume and surface resistivity. This new substrate has great promise for future pulp and paper products.

Acknowledgments

The authors gratefully acknowledge the kind and continued support of the member companies of the Institute of Paper Science and Technology. Portions of this work will serve as partial fulfillment for the Ph.D. degree requirements for D.G.M.

References

1. Quignard, F.; Choplin, A. Cellulose: a new bio-support for aqueous phase catalysis. *Chem. Commun.* **2001**, 21–22.
2. Basso, M. C.; Cerrella, E. G.; Cukierman, A. L. Lignocellulosic materials as potential biosorbents of trace toxic metals from wastewater. *Ind. Eng. Chem. Res.* **2002**, *41*(15), 3580–3585.
3. Ghosh, P., Gangopadhyay, R. Photofunctionalization of cellulose and lignocellulose fibres using photoactive organic acids. *Eur. Poly. J.* **2000**, 36(3), 625–634.

Chapter 14

FIBER MODIFICATION *VIA* DIELECTRIC-BARRIER DISCHARGE
Theory and practical applications to lignocellulosic fibers

L. C. Vander Wielen and A. J. Ragauskas

Institute of Paper Science and Technology, 500 Tenth Street NW, Atlanta, GA 30332-0620, U.S.A.

14.1 Introduction

Given global emphasis on environmental awareness, the demand for innovative technologies that apply green chemistries to renewable resources for the 21st century will continue to grow [1]. The application of cold plasmas to wood fibers offers means to alter the surface chemistry and physical properties of one of the world's most abundant renewables. Plasma, the fourth state of matter, can be generated by a variety of methods including high heat conditions, the application of electromagnetic waves at radio and microwave frequencies in a vacuum, or when electrons in an electric current gain energy in amounts sufficient to separate gaseous atoms and molecules causing ionization.

The invention of the dielectric-barrier discharge device, which applies an electric current for plasma formation, is attributed to Werner von Siemens, who introduced the concept of dielectric-barrier discharge treatment of air for ozone generation in 1857 [2]. Today, dielectric-barrier discharges continue to be used for ozone generation [3] but have found applications in pollution control [4], silent discharge CO_2 lasers and ultraviolet excimer lamps [5], plasma displays [6], and the surface treatment of polymers [7]. The purpose of this review is to provide an overview of current and relevant research into the potential application of dielectric-barrier discharge to the surface modification of lignocellulosic fibers. Impacts of treatment in terms of the surface chemistry, physical strength properties, and water affinity of lignocellulosic fibers are discussed.

J. V. Edwards et al. (eds.), Modified Fibers with Medical and Specialty Applications, 215–229.
© 2006 *Springer. Printed in the Netherlands.*

14.2 Dielectric-barrier discharge treatment

Dielectric-barrier discharge treatment involves a metal high-voltage treatment electrode, a ground electrode, and an insulating material covering at least one electrode to allow for the formation of glow discharges at atmospheric pressure, as opposed to the formation of one or several localized streamers [8]. However, the dielectric-barrier discharge is not without streamers as it contains numerous micro-discharges that exist for nanoseconds [9]. These micro-discharges increase in number as treatment power or time is increased while the properties of individual micro-discharges remain constant [5]. Surface treatment occurs when 10–30 kV are applied across 1–3 mm gaps between treatment electrodes, ionizing the air in the gap that appears as a visible violet–blue corona [5, 10–14]. Figure 14.1 typifies the surface treater used in our research studies, which is representative of industrial dielectric-barrier discharge configurations designed for the surface treatment of moving webs.

Dielectric-barrier discharge treatment applies a low temperature plasma of approximately 27–35 °C [13, 14] to materials. In our laboratory, the application of dielectric-barrier discharge treatments to paper sheets formed from ligno-cellulosic fibers in a 1.5-mm treatment gap caused sheet temperatures to reach as much as 40 °C at high treatment levels when samples were passed repeatedly across treatment electrodes at 5 m/min as measured by Thermography [15]. A maximum temperature of 36–40 °C was confirmed using Cole Parmer temperature indicator strips [15]. Cold plasmas for surface treatment can also be generated using radio frequency waves and microwaves, which require vacuum conditions. In contrast, dielectric-barrier discharges can be performed at atmospheric conditions, making dielectric-barrier discharge technologies advantageous from a practical standpoint [16].

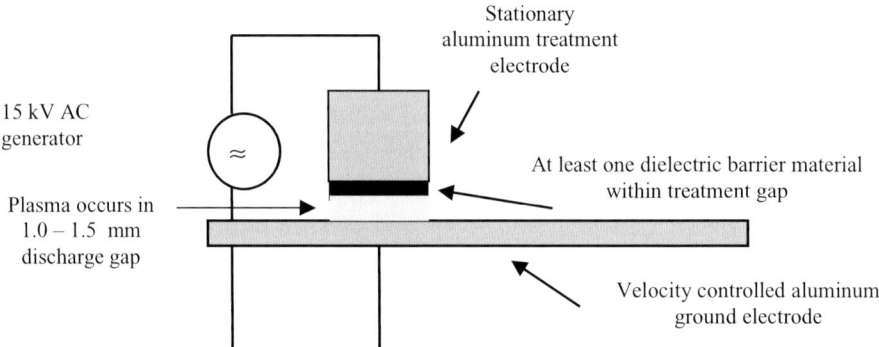

Figure 14.1. Schematic for a dielectric-barrier discharge treater

An atmospheric dielectric-barrier discharge plasma consists of photons [17], electrons [2, 5], free radicals [18], ions [2, 19], and excited molecules [20]. A major by-product of dielectric-barrier discharge is the production of ozone gas in the air [2]. When the high-energy electrons reach the surface of substrates, they have energies sufficient to break molecular bonds resulting in the creation of ions, free radicals, and other species on treated surfaces [10]. This leads to changes in surface energy, surface oxidation, cross-linking, broken bonds, disruption of surface layers, surface cleaning, and creation of free radicals on the surface of treated materials [10, 12, 21–23].

14.2.1 Physical and chemical modification of lignocellulosic fibers

Throughout the literature, dielectric-barrier discharge-initiated surface treatment is associated with increased surface energy and wettability as evidenced by the decreased contact angles of water on polymeric and lignocellulosic surfaces [23–26]. In our laboratory, the contact angle of nano-pure water on the surface of polyester film decreased from 65 ° to 30 ° with 0.056 kW/m²/min applied dielectric-barrier discharge treatment [15]. Decreased contact angles of water on the surface of cellophane with increased dielectric-barrier discharge treatment have been reported [25]. In our studies, the treatment of unbleached thermomechanical pulp fibers at low treatment dosages led to decreases in contact angle of approximately 12%; however, this effect diminished as treatment levels were increased (Table 14.1) [27].

Experimental difficulties measuring contact angles on rough, fibrous, and absorbent surfaces make inverse gas chromatography desirable for examining changes in the surface energy of lignocellulosic fibers [28–32]. Research studies reported that the dispersive surface energy of α-cellulose powder increased from 31.9 to 46.3 mJ/m² with corona discharge treatment [33]. However, the dispersive surface energies for dielectric-barrier discharge-treated bleached kraft and unbleached mechanical pulp fibers increase by 27.5% and 8.7%, respectively,

Table 14.1. Contact angles for dielectric-barrier discharge-treated fibers [27]

Treatment intensity (kW/m²/min)	Contact angle (°)
0.0	58.2
1.5	40.8
3.3	45.5
6.0	49.5
9.3	49.2

Table 14.2. Change in ratio of O/C detected *via* ESCA for dielectric-barrier
discharge-treated fibers

Substrate	Percentage increase in O/C ratio due to surface treatment	Reference
Hardwood α-cellulose powder	10.1	[33]
Pine thermomechanical pulp	10.3	[15]
Spruce thermomechanical pulp (acetone extracted)	9.5	[27]
Bleached kraft pulp (acetone extracted)	2.8	[27]
Newsprint	20.5	[40]

with 0.12 kW/m^2/min applied dielectric-barrier discharge treatment [27]. As treatment intensity is further increased, the dispersive surface energies decrease until at or below those of untreated samples. This trend agrees with trends seen when the contact angle (Table 14.1) was investigated [27].

Electron spectroscopy for chemical analysis (ESCA) has been widely applied for examining the surface chemistry of lignocellulosic fibers [34–39]. The determination of surface O/C ratios *via* ESCA indicate that lignocellulosic materials are typically oxidized by dielectric-barrier discharge treatment (Table 14.2) [27, 33, 40].

ESCA studies regarding functional groups suggest that the measured surface carboxylic acid content of purified hardwood α-cellulose changes from 0% to 4.7% with dielectric-barrier discharge treatment while aldehyde groups increase from 12.4% to 19.9% [41]. Another study revealed increases in carboxylic acids of 0.04% in thermomechanical pulp sheets and 0.02% on Whatman filter paper, with respective increases of 0.58% and 0.98% among ethanol extracts that were attributed to an oxidative depolymerization and degradation process [21]. In our studies, an increase in surface carboxylic acids on bleached kraft fibers of approximately 45% occurred with 0.12 kW/m^2/min dielectric-barrier discharge treatment, with decreases of only 20% detected upon acetone extraction of treated samples due to the removal of degraded materials as detected *via* ESCA and surface titration methods (Table 14.2) [27, 42]. For dielectric-barrier discharge-treated unbleached thermomechanical pulp fibers, a 65% increase in carboxylic acid groups was measured before and a 23% net increase was measured after acetone extraction to remove degraded materials [42]. The spikes in surface acids at low treatment levels show a trend that is strikingly similar to the spike in dispersive surface energy seen at 0.12 kW/m^2/min [27]. In addition, the aldehyde content of thermomechanical pulp fibers was approximately doubled with dielectric-barrier discharge treatment [27, 42].

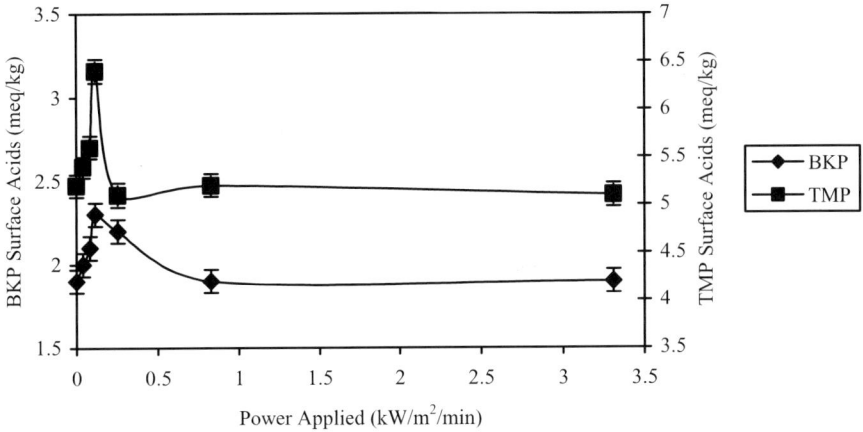

Figure 14.2. Impact of dielectric-barrier discharge treatment upon surface acid content of fibers measured by polyelectrolye titration [27, 42]

Wet chemical methods using methylene blue dye, methylene violet dye, and Schiff's reagent followed by spectrophotometry, and the measurement of colorimetric nitrogen content were used to explore carboxyl and aldehyde groups. These methods indicated no increase carboxylic acid groups, and possibly a decrease, but did detect an increase in aldehydes [43]. The application of bulk conductometric titrations to fibers failed to detect changes in the carboxylic acid content of treated pulps [42]. The dye and conductometric titration methods measure the total acid groups, as opposed to surface acids. These methods are relatively insensitive to changes in the fiber surface acid groups since the large number of acid groups throughout the fiber may make it difficult to detect the impact of surface treatment. Nonetheless, detectable increases in surface acids were observed (Figure 14.2) when employing dielectric-barrier discharge treatment at low energy treatment dosages [27, 42].

Along with the surface chemistry effects of dielectric-barrier discharge on lignocellulosic fibers, several reports have documented changes to physical properties. The coefficient of friction of newsprint has been shown to increase with increased dielectric-barrier discharge treatment [40] while treated bleached kraft and unbleached thermomechanical pulps exhibited increases in the static coefficient of friction at low treatment levels, which reverse at high treatment levels (9.3 kW/m²/min), as indicated in Table 14.3 [27]. Atomic force microscopy (AFM) images revealed a rough, fibrillar appearance at low dielectric-barrier discharge treatment, and a smoothing of the fiber surface at high dielectric-barrier discharge treatment, which was quantified using AFM to measure the root-mean-square (RMS) roughness [27]. The RMS roughness of bleached kraft and unbleached thermomechanical pulp fibers increased by

Table 14.3. Impact of dielectric-barrier discharge treatment on the static coefficient of friction (COF) and viscosity of lignocellulosic fibers [27]

Surface treatment (kW/m²/min)	COF for fully bleached kraft pulp	COF for unbleached thermomechanical pulp	Viscosity (cP) of bleached kraft
0.0	0.71	0.54	13.73
0.1	0.80	0.63	13.01
3.3	0.81	0.71	12.34
6.0	—	—	5.96
9.3	0.72	0.48	4.94

42% and 31%, respectively, at low treatment levels, then became smoother until roughness levels were similar to those of untreated samples as treatment intensity was increased [27]. Physical changes to cellulosic fibers were also reported in terms of pulp viscosity [27], which decreases as dielectric-barrier discharge treatment is increased (Table 14.3).

Damage to over-treated fibers, along with other property changes, can be seen when fibers are over-treated with dielectric-barrier discharge. Scanning electron microscopy (SEM) shows fibril damage and pin-holing at increased treatment levels [48]. In our laboratory, pin-holing of southern pine thermomechanical pulp sheets was seen at treatment levels at 0.13 kW/m²/min or greater [15], as illustrated by Figure 14.3. Recent studies report treatment of fibrous

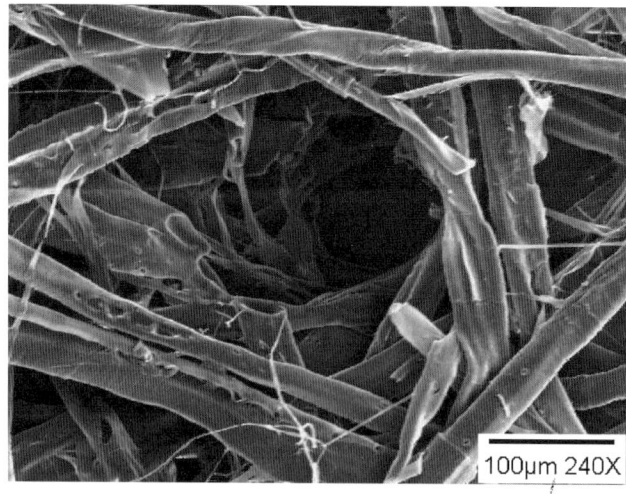

100µm 240X

Figure 14.3. Pin-hole formed due to arcing in dielectric-barrier discharge system due to over-treatment of unbleached southern pine thermomechanical pulp fibers

Table 14.4. Change in brightness of dielectric-barrier discharge-treated fibers [15]

Surface treatment level, bleached kraft pulp (kW/m²/min)	Brightness (%)	Surface treatment level, unbleached thermomechanical pulp (kW/m²/min)	Brightness (%)
0.0	84.9	0	51.5
0.1	85.2	1.49	50.8
0.233	85.2	3.31	49.8
3.3	85.9	5.96	49.1
9.3	86.0	9.27	46.6

non-wovens involving the coverage of both electrodes by dielectric materials, rather than only one electrode, results in diminished pin-holing [44].

In addition, the brightness of bleached kraft fibers increases with dielectric-barrier discharge treatment while the brightness of unbleached thermomechanical pulp fibers decreases (Table 14.4) [15]. This is likely due to the surface cleaning of bleached kraft fibers [27] versus the oxidation of lignin on the surface of unbleached fibers to form chromophores [45].

14.2.2 Dielectric-barrier discharge induced grafting onto lignocellulosic fibers

An alternative application of dielectric-barrier discharge is to initiate grafting of materials onto lignocellulosics. Early studies into this avenue of research employed a two-step methodology [46]. In the first stage, a dielectric-barrier discharge was applied under reduced pressure (0.1mmHg) in the presence of nitrogen or air. The samples were subsequently immersed into an aqueous solution of ethyl acrylate. Untreated films showed no grafting, whereas treated films showed incorporation of up to 110% by weight at optimal conditions [46]. Dielectric-discharge treatment was also used to prime a fibrous cellulose sheet (Whatman filter paper) for grafting [47, 48]. After dielectric-barrier discharge treatment, styrene was grafted to the paper's surface in methanol. Optimal grafting occurred at a styrene concentration of 50%, with the bulk of grafting occurring in the first 30 s [48].

Our research group recently reported the dielectric-barrier discharge-initiated *in-situ* grafting of acryl amide [49] and maleic acid [50] onto lignocellulosic fibers. ToF-SIMS, SEM, elemental analysis, and titration methods showed that increased dielectric-barrier discharge treatment intensity results in increased grafting, that grafting occurs more abundantly on fully bleached kraft fibers than onto unbleached mechanical pulp fibers, and that grafting may be performed *in situ* rather than in a two-step process [49, 50]. When comparing

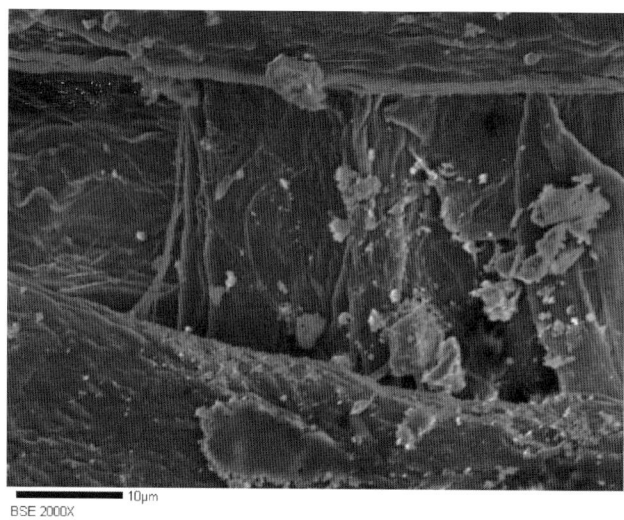

BSE 2000X

Figure 14.4. SEM micrograph of acryl amide grafted onto bleached kraft pulp fibers at 1.2
kW/m²/min treatment taken at 2000×

the *in-situ* and two-step grafting methods, the yield for the *in-situ* grafting is lower [46–50]; however, *in-situ* grafting can be performed quickly in a continuous web-fed process. Both methods result in grafting that appears to occur unevenly across the fiber surface [46–50]. The uneven grafting seen in our studies is described in Figure 14.4.

Given a high enough treatment dosage and sufficient monomer, the *in-situ* grafting process results in the formation of a composite material in which the grafted material coats the surface of the fibers [49] as described in Figure 14.5.

Two reaction mechanisms for dielectric-barrier discharge-initiated grafting onto cellulose in solution have been proposed [46, 48]. The "trapped radical" mechanism involves grafting to cellulose when monomers diffuse to react with trapped radicals within the fiber that become accessible in solution due to fiber swelling [46]. The other mechanism proposed is the formation of peroxides at the fiber surface, which would break down in solution when heat is applied, forming radicals in solution and on the fiber surface [48]. Bataille *et al.* [47] attributed the bimodal molecular weight distribution found when styrene was grafted onto cellulose film to a combination of both of these mechanisms. However, Sakata and Goring [46] found that when grafting ethyl acrylate to cellulose films, little to no homopolymer formed, and ferrous ion, which is known to catalyze the decomposition of peroxy species into radicals, actually inhibited grafting. For these reasons, they concluded that it was trapped radicals, rather than the formation of peroxides at the surface, which acted as grafting

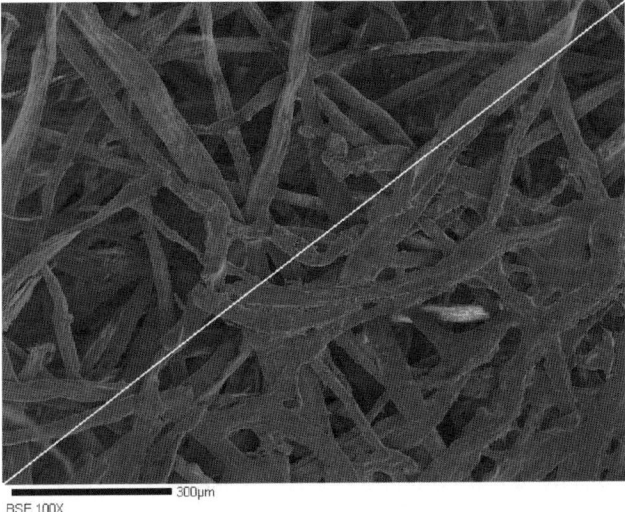

Figure 14.5. SEM micrograph of acryl amide grafted onto bleached kraft pulp fibers at 4.7 kW/m²/min. Grafted (bottom right) versus unmodified fiber (upper left) are compared

initiators. In the case of *in-situ* grafting, the reactions are tentatively attributed to a free-radical mechanism [49].

14.2.3 Bonding of lignocellulosic fibers

The application of dielectric-barrier discharge to lignocellulosic materials for improved dry-strength performance was reported (1967) and patented (1969) by Goring [12, 51]. Pressing together layers of surface-treated cellulose acetate films or sheets formed from bleached hardwood and softwood kraft fibers increases the bond strength between laminates [11, 12, 43, 52]. The adhesion between treated laminates of wood, cellulose films, or paper with treated synthetic polymers can also be improved *via* dielectric-barrier discharge [11, 53]. Chemical additives impact this process, as the benefits to adhesion between paper and polyethylene upon surface treatment are enhanced in the presence of starch [54] while calcium carbonate fillers reduce adhesion [55]. Dielectric-barrier discharge is also applied to improving the adhesion of glues [56], paints [24], and inks [57] to lignocellulosic fibers. For example, the adhesion of toner ink to paper [57] can be enhanced by dielectric discharge in a photocopy machine.

When bleached kraft or unbleached thermomechanical pulp fibers were dielectric-barrier discharge treated over a wide range of treatment intensities (0–9.3 kW/m²/min) and subsequently formed into sheets, the only significant improvements to bonding detected were increases in the wet-tensile index.

Table 14.5. Change in the wet tensile of dielectric-barrier discharge-treated fibers [59, 61]

Surface treatment (kW/m²/min)	Wet-tensile index-bleached kraft (Nm/g)	Wet-tensile index thermomechanical pulp (Nm/g)	Wet-tensile index thermomechanical pulp (kNm/g)
0	0.53	0.67	0.045
3.3	0.71	0.86	0.073
6.0	0.83	1.17	0.090
9.3	0.91	2.53	0.111

Sheets formed from surface-treated bleached kraft and unbleached thermomechanical pulp sheets showed wet-strength increases of 20% and 60%, respectively [58, 59]. The dry-strength properties such as tensile, tear, and z-direction tensile remained unchanged. When fibers were grafted with monomers, such as acryl amide and maleic acid, followed by formation into sheets, similar results were observed [50, 58].

It was reported that treated sheets formed from bleached kraft pulp fibers with and without added polyacrylamide showed increases in breaking length (dry-tensile strength) at treatment levels up to 3.2 kW/m²/min, which diminished above this treatment level [60]. In our studies, the surface treatment of formed kraft and thermomechanical pulp sheets at treatment levels ranging up to 9.3 kW/m²/min caused no statistically significant changes to dry-strength properties, including tensile, tear, and z-direction tensile [58, 59]. However, the wet-tensile index of bleached kraft and unbleached thermomechanical pulp sheets increased with increased dielectric-barrier discharge treatment (Table 14.5) whether or not polymers were added [58, 59]. Since the magnitude of increases seen without polymer were similar to those seen when polymer was added, the data reported in Table 14.5 reports only data acquired when testing fibrous sheets not containing polymer. Wet-stiffening of fibers is also seen with increased dielectric-barrier discharge treatment [61].

14.2.4 Water absorption of lignocellulosic fibers

The water affinity properties of lignocellulosics are also impacted by plasma treatment. Modification by means of carbon tetrafluoride plasmas [62], the plasma deposition of hexamethyldisiloxane [63], and the application of other water repellent additives provides hydrophobic fiber surfaces. The atmospheric dielectric-barrier discharge treatment of wood causes water uptake to increase while dielectric-barrier discharge treatments in methane and acetylene render the surface of wood more hydrophobic [64]. The volumetric swelling

Table 14.6. Impact of dielectric-barrier discharge on the water affinity properties of fibers [15, 27, 59]

Surface treatment level (kW/m²/min)	TAPPI WRV at 900g	WRV at 225g	Percentage water uptake	Percentage change in linear dimensional stability	Vertical water wicking (cm/s²)
Bleached kraft pulp					
0.0	0.93	2.93	0	0	0.66
0.1	0.97	3.04	11.50	−2.56	0.66
3.3	0.89	2.76	−2.59	17.59	0.69
9.3	0.84	2.66	−34.24	49.75	0.69
Unbleached thermomechanical pulp					
0.0	1.15	4.0	0	0	0.76
0.1	1.15	3.7	13.84	−16.67	0.92
3.3	1.14	3.4	−4.94	25.00	0.89
9.3	1.13	3.1	−57.77	82.30	0.85

of cellophane films increases by up to 80% with up to 5.0 min corona discharge treatment [25]. Both the water absorption of ply-bonded paper formed from corona-treated bleached kraft hand sheets and the moisture absorption of corona-treated bleached kraft pulp were lower than that of untreated reference samples [43].

Our studies indicated both hydrophilic and hydrophobic behaviors among dielectric-barrier discharge-treated lignocellulosic fibers depending upon treatment intensity (Table 14.6) [27, 59]. The water retention value (WRV), which is indicative of fiber swelling, was tested by both Tappi Useful Method 256 [65], which provides centrifugation at 900g, and an additional lower acceleration (225g) method [15]. The WRV tests, change in linear dimensional stability, and percent water uptake each indicated increases in the water affinity properties of bleached kraft pulp at low treatment levels, which diminished with increased treatment [27, 59]. However, water-wicking studies detected no statistically significant changes in the vertical wicking of bleached kraft fibers. The 225g WRV test, analysis of change in linear dimensional stability, change in percent water uptake, and wicking studies performed using thermomechanical pulp fibers also indicated an increase in water affinity properties at low dielectric-barrier discharge treatment levels, which decrease with increased surface treatment [27, 59]. The spikes in water affinity at low treatment levels show trends that are strikingly similar to spikes in surface acids, the dispersive surface energy, and surface roughness previously discussed. These properties also diminished with increased dielectric-barrier discharge treatment intensity.

14.3 Conclusion

Surface treatment *via* atmospheric dielectric-barrier discharge has shown great potential for modifying a series of chemical and physical properties of lignocellulosic materials. The observed wet-strength benefits realized when lignocellulosic fibers are dielectric-barrier discharge treated have obvious applications in the pulp and paper. In addition, the ability to graft acrylic derivatives onto fibers provides a tremendous opportunity for the generation of biocomposites. A notable feature of these treatments is they can be potentially performed in a continuous process without requiring vacuum conditions or special solvents. In addition, this process has been shown to tailor the surface topochemistry of lignocellulosic fibers by simply adjusting treatment dosages with and without chemical additives. Further research and development of dielectric-barrier discharge applications to lignocellulosic fibers will undoubtedly be developed, as it provides a green method for altering the surface chemistry of the world's most abundant renewable resource.

Acknowledgments

The authors wish to acknowledge the support of the member companies of the Institute of Paper Science and Technology at the Georgia Institute of Technology. Portions of this work are being used by Lorraine C. Vander Wielen as partial fulfillment of the requirements for graduation from the Institute of Paper Science and Technology, 500 Tenth Street, NW, Atlanta, GA 30332-0620, U.S.A.

References

1. Singh, S. K.; Gross, R. A. Overview: Introduction to polysaccharides, agroproteins, and poly(amino acids). In: Gross, R. A.; Scholz, C. (Eds.) *Biopolymers from Polysaccharides and Agroproteins, ACS Symposium Series 786*. American Chemical Society, Washington DC, **2001**, 2–40.
2. Kogelschatz, U.; Eliasson, B.; Egli, W. From ozone generators to flat television screens: History and future potential use of dielectric-barrier discharges. 14th International Symposium on Plasma Chemistry, Praha, Czech Republic, **1999**.
3. Kuraica, M. M.; Obradovic, B. M.; Manojlovic, D.; Ostojic, D. R.; Puric, J. New type of coaxial dielectric-barrier-discharge used as ozonized water generator. *Adv. Appl. Plasma Sci.* **2003**, *4*, 415–418.
4. Ighigeanu, D.; Martin, D.; Macarie, R.; Zissulescu, E.; Calinescu, I.; Iovu, H.; Cirstea, E.; Craciun, G.; Ighigeanu, A. Air pollution control by DC, pulse and microwave discharges. *J. Environ. Prot. Ecol.* **2003**, *4*, 525–534.
5. Kogelschatz, U. Dielectric-barrier discharges: Their history, discharge physics, and industrial applications. *Plasma Chem. Plasma Process.* **2003**, *23*, 1–46.

6. Boeuf, J. P.; Pitchford, L. C. Calculated characteristics of an ac plasma display panel cell. *IEEE Trans. Plasma Sci.* **1996**, *24*, 95–96.
7. Sun, C. Q.; Zhang, D.; Wadsworth, L. C. Corona treatment of polyolefin films—A review. *Adv. Polym. Technol.* **1999**, *18*, 171–180.
8. Kogelschatz, U. Filamentary, patterned, and diffuse barrier discharges. *IEEE T Plasma Sci.* **2002**, *30*(4 Pt. 1), 1400–1408.
9. Eliasson, B.; Hirth, M.; Kogelschatz, U. Ozone synthesis from oxygen in dielectric barrier discharges. *J. Phys. D. Appl. Phys.* **1987**, *0*, 1421–1437.
10. Cramm, R. H.; Bibee, D. V. The theory and practice of corona treatment for improving adhesion. *Tappi J.* **1982**, *65*, 75–78.
11. Kim, C. Y.; Suranyi, G.; Goring, D. A. I. Corona induced bonding of synthetic polymers to cellulose. *J. Polym. Sci. C* **1970**, *30*, 533–542.
12. Goring, D. A. I. Surface modification of cellulose in a corona discharge. *Pulp Pap. Mag. Can.* **1967**, *68*(8), T372–T376.
13. Raizer, Y. P. *Gas Discharge Physics.* Springer-Verlag, Berlin, **1991**, 1–6 (based on original Russian edition, Fizika Gazovogo Razryada).
14. Rehn, P.; Wolkenhauer, M.; Bente, M.; Förster, S.; Viöl, W. Wood surface modification in dielectric barrier discharges at atmospheric pressure. *Surf. Coat. Technol.* **2003**, 174–175, 515–518.
15. Vander Wielen, L. C. Dielectric Barrier Discharge-Initiated Fiber Modification. Dissertation, Institute of Paper Science and Technology, Atlanta, GA, **2004**.
16. Carlsson, G.; Ström, G. Water sorption and surface composition of untreated or oxygen plasma-treated chemical pulps. *Nord. Pulp Pap. Res. J.* **1995**, *10*, 17–23, 32.
17. Shen, B.; Yu, M. Y.; Wang, X. Photon–photon scattering in a plasma channel. *Phys. Plasmas* **2003**, *10*, 4570–4571.
18. Naidis, G. V. Modeling of plasma chemical processes in pulsed corona discharges. *J. Phys. D. Appl. Phys.* **1997**, *30*, 1214–1218.
19. Shahin, M. M. Nature of charge carriers in negative coronas. *Appl. Opt. Suppl. Electrophotogr.* **1969**, *3*, 106–110.
20. Denes, F.; Simionescu, C. I. Use of plasma chemistry in the synthesis and modification of natural macromolecular compounds. *Cell. Chem. Technol.* **1980**, *14*, 285–316.
21. Nishiyama, S.; Funato, N.; Sawatari, A. Analysis of functional groups formed on the corona treated cellulose fiber sheet surface by means of chemical modification in gas phase ESCA technique. *Sen-I Gakkaishi* **1993**, *49*, 73–82.
22. Goossens, O.; Dekempeneer, E.; Vangeneugden, D.; Van de Leest, R.; Leys, C. Application of atmospheric pressure dielectric barrier discharges in deposition, cleaning, and activation. *Surf. Coat. Technol.* **2001**, *142–144*, 474–481.
23. Lawson, D.; Greig, S. Bare roll treaters vs. covered roll treaters. Polymers, Laminations, and Coating Conference, **1997**, 681–693.
24. Back, E. L.; Danielsson, S. Oxidative activation of wood and paper surfaces for bonding and for paint adhesion. *Nord. Pulp Pap. Res. J.* **1987**, 53–62.
25. Brown, P. F.; Swanson, J. W. Wetting properties of cellulose treated in a corona discharge. *Tappi J.* **1971**, *54*, 2012–2018.
26. Bezigian, T. The effect of corona discharge on polymer films. *Tappi J.* **1992**, *75*, 139–141.
27. Vander Wielen, L. C.; Elder, T.; Raguaskas, A. J. Analysis of the topochemistry of dielectric-barrier discharge treated cellulosic fibers. *Cellulose.* **2005**, *12*(2), 185–196.

28. Santos, J. M. R. C. A.; Gil, M. H.; Portugal, A.; Guthrie, J. T. Characterization of the surface of a cellulosic multi-purpose office paper by inverse gas chromatography. *Cellulose* **2001**, *8*, 217–224.

29. Jacob, P. N.; Berg, J. C. Acid–base surface energy characterization of microcrystalline cellulose and two wood pulp fiber types using inverse gas chromatography. *Langmuir* **1994**, *10*, 3086–3093.

30. Garnier, G.; Glasser, W. G. Measurement of the surface free energy of amorphous cellulose by alkane adsorption: A critical evaluation of inverse gas chromatography (IGC). *J. Adhes.* **1994**, *46*, 165–180.

31. Liu, F. P.; Rials, T. G.; Simonson, J. Relationship of wood surface energy to surface composition. *Langmuir* **1998**, *14*, 536–541.

32. Felix, J. M.; Gatenholm, P. Characterization of cellulose fibers using inverse gas chromatography. *Nord. Pulp Pap. Res. J.* **1993**, *8*, 200–203.

33. Belgacem, M. N.; Blayo, A.; Gandini, A. Surface characterization of polysaccharides, lignins, printing ink pigments, and ink fillers by inverse gas chromatography. *J. Colloid Interface Sci.* **1996**, *182*, 431–436.

34. Dorris, G. M.; Gray, D. G. The surface analysis of paper and wood fibers by ESCA (electron spectroscopy for chemical analysis). I. Application to cellulose and lignin. *Cell. Chem. Technol.* **1978**, *12*, 9–23.

35. Dorris, G. M.; Gray, D. G. The surface analysis of paper and wood fibers by ESCA. II. Surface composition of mechanical pulps. *Cell. Chem. Technol.* **1978**, *12*, 721–734.

36. Koljonen, K.; Österberg, M.; Johansson, L.-S.; Stenius, P. Surface chemistry and morphology of different mechanical pulps determined by ESCA and AFM. *Colloids Surfaces A. Physicochem. Eng. Aspects* **2003**, *228*, 143–158.

37. Hulten, A. H.; Paulsson, M. Surface characterization of unbleached and oxygen delignified kraft pulp fibers. *J. Wood Chem. Technol.* **2003**, *23*, 31–46.

38. Gellerstedt, F.; Gatenholm, P. Surface properties of lignocellulosic fibers bearing carboxylic groups. *Cellulose* **1991**, *6*, 103–121.

39. Laine, J.; Stenius, P.; Carlsson, G.; Stroem, G. The effect of elemental chlorine-free (ECF) and totally chlorine-free (TCF) bleaching on the surface chemical composition of kraft pulp as determined by ESCA. *Nord. Pulp Pap. Res. J.* **1996**, *11*, 201–210.

40. Gurnagul, N.; Ouchi, M. D.; Dunlop-Jones, N.; Sparkes, D. G.; Wearing, J. T. Coefficient of friction of paper. *J. Appl. Polym. Sci. 46*, 805–814.

41. Belgacem, M. N.; Czeremuszkin, G.; Sapieha, S. Surface characterization of cellulose fibers by XPS and inverse gas chromatography. *Cellulose* **1995**, *2*, 145–157.

42. Vander Wielen, L. C.; Raguaskas, A. J. Dielectric discharge: A concatenated approach to fiber modification. Proceedings of the 12th International Symposium on Wood and Pulping Chemistry, Vol. 1, Madison, WI, **2003**, 373–376.

43. Sakata, I.; Morita, M.; Furuichi, H.; Kawaguchi, Y. Improvement of ply-bond strength of paperboard by corona treatment. *J. Appl. Polym. Sci.* **1991**, *42*, 2099–2104.

44. Borcia, G.; Anderson, C. A.; Brown, N. M. D. Dielectric barrier discharge for surface treatment: Application to selected polymers in film and fiber form. *Plasma Sci. Technol.* **2003**, *12*, 335–344.

45. Bukovsky, V.; Trnkova, M. The influence of secondary chromophores on the light induced oxidation of paper. Part II: The influence of oxidation of paper. *Restaurator* **2003**, *24*, 118–132.

46. Sakata, I.; Goring, D. A. I. Corona-induced graft polymerization of ethyl acrylate onto cellulose film. *J. Appl. Polym. Sci.* **1976**, *20*, 573–579.

47. Bataille, P.; Dufourd, M.; Sapieha, S. Copolymerization of styrene on to cellulose activated by corona. *Polym. Int.* **1994**, *34*, 387–391.
48. Bataille, P.; Dufourd, M.; Sapieha, S. Graft polymerization of styrene onto cellulose by corona discharge. *Polym. Preprints* **1991**, *32*, 559–560.
49. Vander Wielen, L. C.; Raguaskas, A. J. Grafting of acrylamide onto lignocellulosic fibers via dielectric-barrier discharge. *Eur. Polym. J.* **2004**, *40*, 477–482.
50. Vander Wielen, L. C.; Ragauskas, A. J. Dielectric-barrier discharge treatment: A palmary approach to fiber modification. American Institute of Chemical Engineers National Meeting, San Francisco, CA, **2003**, 489a.
51. Goring, D. A. I. Surface modification of cellulose. Canadian Patent 8304689, Pulp and Paper Research Institute of Canada, **1969**.
52. Kim, C. Y.; Goring, D. A. I. Corona induced bonding of synthetic polymers to wood. *Pulp Pap. Mag. Can.* **1971**, *82*, 93–96.
53. Kemppi, A. Studies on adhesion between paper and low-density polyethylene. 1 Influence of the natural components in paper. *Paperi ja Puu* **1996**, *78*, 610–617.
54. Kemppi, A. Adhesion between paper and low density polyethylene. 2. The influence of starch. *Paperi ja Puu* **1997**, *79*, 178–185.
55. Kemppi, A. Adhesion between paper and low density polyethylene. 3. The influence of fillers. *Paperi ja Puu* **1997**, *79*, 330–338.
56. Back, E. L. Oxidative activation of wood surfaces for glue bonding. *Forest Products J.* **1991**, *41*, 30–36.
57. Berkes, J. S.; Bonsignore, F. J. Xerox Corp. Process for obtaining a very high transfer efficiency from intermediate to paper. United States Patent. No. 5119140, **1992**.
58. Vander Wielen, L. C.; Page, D. H.; Ragauskas, A. J. Impact of dielectric-barrier discharge on bonding. 2003 International Paper Physics Conference Pre-prints, PAPTAC, Victoria, British Columbia, Canada, **2003**, 347–349.
59. Vander Wielen, L. C.; Page, D. H.; Ragauskas, A. J. Enhanced wet-tensile paper properties via dielectric-barrier discharge treatment. *Holzforschung* **2005**, *59*, 65–71.
60. Nishimura, J.; Nakao, T.; Uehara, T.; Yano, S. Improvement of paperboard mechanical properties by corona-discharge treatment. *Tappi J.* **1990**, *73*, 275–276.
61. Vander Wielen, L. C.; Ragauskas, A. J. Wet-stiffening of TMP and kraft fibers via dielectric-barrier discharge treatment. *Nord. Pulp Pap. Res. J.* **2004**, *19*, 384–385.
62. Young, R. A.; Denes, F.; Hua, Z. Q.; Sabharwal, H.; Nielsen, L. Cold plasma modification of lignocellulosic material. International Symposium on Wood and Pulping Chemistry, Helsinki, Finland, **1995**, 637–644.
63. Denes, A. R.; Tshabalala, M. A.; Rowell, R.; Denes, F.; Young, R. A. Hexamethyldisiloxane-plasma coating of wood surfaces for creating water repellent characteristics. *Holzforschung* **1999**, *53*, 318–326.
64. Rehn, P.; Viöl, W. Dielectric-barrier discharge treatments at atmospheric pressure for wood surface modification. *Holz als Roh-und Werkstoff* **2003**, *61*, 145–150.
65. Tappi useful method 256. *TAPPI Useful Methods*, Vol. 1991. Atlanta, GA, 54–56, **1991**.

INDEX

1,2,3,4-butanetetracarboxylic
 (BTCA) 160
application 163
1,2,3,4-cyclopentanetetracarboxylic
 acid 160
^{31}P NMR 174
8-Br-AMP 40
absorbency 11, 16, 19, 21, 24, 31, 199
active cotton wound dressings 26
 performance in chronic wounds 29
 prototype design of 24
active cotton-based wound dressings,
 design 25, 26
ADMH 86–88
 -treated fibers 85
adsorption isotherms 50
aglycon 129, 130, 132
albumin 5, 29, 30, 49, 51, 52, 55, 57–60,
 62, 63, 107, 110, 112, 151
 absorbance 55
 adsorption 53
 detection 53
alginates 11, 15, 16, 19–22, 24
alkali treatment 181
alkaline hydrolysis 98–100, 107, 109,
 161, 167, 172, 178
α-hydroxyalkylamides 160
amide 82–85, 160–164, 166–171, 178,
 185, 187, 188, 221–224
amine 40, 82, 83, 100, 111, 167–169,
 184

amino acid sequence 3
aminolysis 98, 100, 101, 109
AN69 36, 37, 40, 43, 44
animal fibers 12
anionic dyes 75, 100
anti-bacterial fibers, for producing
 clothing and filters 1
anti-HIV 128
antimicrobial activity 113–115
anti-microbial bioactive polyester
 surface, development 113
antimony III 92
anti-tumor
 activity 126–129, 134–136, 138,
 139
 agents 126
 effect 126
anti-viral activity 134, 135, 137, 138
aortic dissection 145, 151, 152
apoptosis 35, 41, 42, 44
Arg-Gly-Asp (RGD) 35, 40, 109, 110
artificial livers 6
artificial polymers 130, 181
Aspartate-102 27
atomic force microscopy (AFM) 200,
 219
average tumoral regression (ATR) 129,
 136

B16 F10 murine melanoma cells 43
bacterial protection 11, 24

231

benzoyl peroxide (BPO) 86–88
bimorphous ceramics 210
biocidal activity 5
biocidal functions 81–83, 85–87
biocidal polymer 81, 85
biocidal textiles 81
biocompatible materials 2, 3
biocompatible scaffolding for tissue
　　regeneration 67
biocompatible surface 107
biodegradability 1, 146
bioerodable implant structures 67
biofiber research 1
BioGlue 151, 152
biological adhesives 6, 145, 154
biological evaluation 134
biomedical applications 2, 52, 67, 71,
　　76, 126
biomedical fibers, principles 6
biopolymers 5, 67, 69, 78
　　biomedical applications 70
blepharoplasty 145
blood anti-coagulant activity 128
blood protein albumin 5
blood proteins 5
bone repair 4, 78
bovine serum albumin (BSA) 52–63,
　　110–115
　　absorbance 55
Brevibacterium imperiale 188
BTCA cross-linked cellulose, strength
　　improvement, by lipases
　　170
BTCA cross-linked fabrics 163, 164,
　　172, 178

cAMP 38–41
Candida rugosa 163, 171
carbohydrate-based wound dressings 11,
　　22, 24
carbohydrates 4, 22, 27, 125, 195, 197,
　　210
carboxylic acid formation 112
carboxylic acids 27, 99, 100, 107, 112,
　　192, 197, 209, 210, 218, 219

carboxymethylated cellulose (CMC) 14,
　　24, 27, 53, 55–61, 63, 128,
　　131–133, 136, 139, 192
cardiovascular grafts 71
cardiovascular surgery
　　biological glues 145, 146, 150–153
　　enbucrilates 145
　　fibrin glues 145, 147–150, 154
catheter cuffs 94, 106, 108
cationic dyes 171–173, 175, 183
cell adhesion 44, 71, 73, 76, 110
cell behavior 6, 35, 41, 53
cell migration 35, 44
cell proliferation 21, 41, 44, 116
cell shape 35, 39
　　observations 36
cell spreading 35, 36, 40
cell-surface integrin receptors 35, 36
cellular adhesion/growth promotion 108
cellular mechanisms 6
cellulases 165, 193, 198, 199
cellulose 2–7, 14, 23, 25–27, 29, 36, 44,
　　49, 50, 52, 58, 62, 78, 83–86, 131,
　　132, 139, 159–166, 169–172,
　　174–179, 191, 192, 198–200,
　　209, 210, 221–223
　　binding domains (CBD) 193, 199, 200
　　diacetate membranes 6
　　fibers 6, 7, 58, 78, 159, 160, 175, 178,
　　　179
　　　chemical modification 159
　　　cross-linked 162, 164, 169, 178
　　　flame retardant finishing 162
　　　new approaches 7
　　　phosphorylated 165
　　　surface modification 159
　　polymers 175
　　　modification 132
cellulose-based nanofibers 78
charcoal cloth 11, 24
chemical additives 223, 226
　　to solid fiber 97
chemical modifications 62, 85, 98, 159,
　　217
chemokines 13, 18

chemotherapy 128, 130
chitosan 11, 24
chromatography 49, 50, 52, 56, 63, 131, 134, 148, 217
Chromolaena odorata 12
chronic dermal ulcers, treatment 23
chronic wound dressings 2, 5, 21, 26, 29, 30, 49
    interactive 21
Cipro-dyed C-EDA segments 113–115
Ciprofloxacin(Cipro) 113–115
citric acid 160
cold plasmas 7, 215, 216
collagen 2–4, 6, 13, 18, 22, 23, 71–73, 76, 107, 109, 156
colorants 96, 164, 166
column chromatography (CMC) 53, 55, 56, 58–60, 62, 131–13, 139
    cotton 55, 59–61, 63
        fibers 57
Comamomas acidovorans 185
condensation reactions 88, 160
connexin 43 organization 38, 39
connexins (Cx) 38, 39
contact layer dressings 13, 16
contraceptives 129
copolymers 73, 74, 76, 128, 186
corona treatments 102, 103, 113, 115
Corynebacterium nitrilophilus 188
cotton 4, 5, 11, 17, 20, 21, 23–29, 49–53, 55–63, 78, 84, 85, 87, 94, 95, 104, 159, 163, 167–174, 176, 177, 195, 196, 198–200
cotton-based interactive wound dressing 26, 49
cotton cellulose phosphorylation 174, 175
cotton fabrics 85, 86, 104, 160, 161, 165, 167, 168, 176–178, 195
    dyeing 176
cotton fiber-protein interactions 49
cotton fibers 5, 29, 51, 52, 54, 56–58, 159, 198, 200
crease-resistance finishing 7, 159, 160, 173

cross-linked cellulose 161, 166, 170
    fibers 162, 178
        dyeability 164
cross-linking 24, 97, 107, 160, 161, 169, 172, 178, 217
cuprophan (CU) 6, 35, 36, 40, 41, 44
curdlan 6, 35, 36, 40, 41, 44, 126, 127
cyanoacrylate 150–154, 156, 157
    adhesives 145, 153
cytokines 11–13, 18, 19, 22, 23, 31
cytoskeletal organization 35

Dacron™ 92, 111–113
deferrioxamine-linked cellulose 23
denier reduction 99
dentures 2, 7
2,3-dialdehyde cellulose (DAC) 131
2,3-dialdehyde carboxymethyl cellulose (DACMC) 131
2,3-dicarboxycellulose (DCC) 131–136
dialysis tubing 1
DIDOX 132, 133, 135, 136, 139
dielectric-barrier discharge 217, 219–221, 223, 225, 226
    device 215
    fiber modification via 215
    treatment 215–221, 224, 225
Digitalis purpurea 130
dimethyloldihydroxyethylenurea (DMDHEU) 160
disinfection 81
DMDMH-treated cellulose 83, 84
drug delivery 21, 74, 76, 78, 128
    devices 67, 70
drug discovery 5, 125, 140
    paradigm 5
drug-carrying polymers 128
drug-cyclic oligomers 128
dye fixation 159, 174–176

elastase 11, 24–27, 29, 49, 51–55, 60–63
    activity 29, 30, 51, 52, 54, 55, 60, 61, 63
    assay 55
    substrate 26, 52

elastic behavior 155, 156
elasticity 2, 11, 21, 24, 70, 71, 73, 148,
150, 155–157
electromagnetic waves 215
electron spectroscopy for chemical
analysis (ESCA) 218
electrophilic cotton 23
electrospinning 67–71, 73, 74, 76, 78
of collagen for scaffolding 71
of collagen for tissue engineering 71
principle 68
electrospraying 68
electrospun 3, 6, 68, 73, 74, 76
coatings 71
fibers 3, 6, 67, 71, 74, 75–78
nanofibers 6, 67
enzymatic hydrolysis 55, 167, 172, 174
duration 166
enzymatic reactions 182, 183
enzyme treatment 163
enzyme-linked fibers 1
Escherichia coli (E. coli) 83, 84, 86–88
ethylene diamine 100
excimer laser treatments 103
excimer lasers 98, 103
exhaustion 98, 166, 175, 176
rates 166
extracellular deposition 18
extracellular matrix proteins 18, 12, 28,
51
extravasation 136

fetal bovine serum (FBS) 35–37, 40
fiber modification, via dielectric-barrier
discharge 215
fiber size 5
fibers 7, 12, 18, 26, 35, 38, 49, 51–53,
56, 58, 59, 63, 67–78, 85, 86, 92,
95, 103, 104, 113, 156, 159–162,
164, 169, 175, 177–179, 181,
182, 185, 186, 188, 191–193,
196, 198–200, 208, 211, 215–226
ADMH-treated 85
enzymatic modification
fibrin glues 145, 147–150, 154

fibrin sealants 6, 146, 147
research 6
fibrinogen 6, 108, 109, 116, 117, 147,
148
fibroblasts 12, 13, 18, 36, 38, 40, 41, 43,
44, 117
fibronectin (FN) 11, 23, 35, 108
fibrophilic dyeing additives 98
fibroplasia 18
fixation 159, 160, 166, 174–176
flame retardant finishing 7
flame treatments 103, 105
fluid balance 11, 24
force of magnetism 210
forest products 7, 191, 193, 200
formaldehyde 83, 150, 151, 160
forskolin 39, 40
frequency-doubled copper vapor laser
(FDCVL) 103
FT-IR spectroscopy 164, 178

galactofuranosyl 127
Gamgee, J. 13
gamma high voltage research 69
Ganoderma lucidum 127
gap junction communication 38, 44
gas chromatography 217
gas-phase processes 105
gelatin 6, 73, 107, 150
gelatin-resorcinol-formaldehyde blue
(GRF) 150, 151
glucans 126, 127
glucose oxidase bleaching 197
glues 145, 156, 223
glycoconjugates 125
glycoscience 125
glucose 3, 127, 129, 130, 164, 1765,
185, 197
oxidase 196, 197
glycosidases 125
grafting 13, 71, 85, 86–88, 97, 98, 105,
106, 194, 210, 221, 222
in-situ 221–223
plasma-induced 105
polymerization 86

granulation tissue 17, 18
growth factor stimulation 22, 30, 31
growth factors 13, 18, 19, 22, 23, 28, 31,
        49, 51, 107, 109, 117
GTP-binding protein RhoA 36

halamine chemistry 81, 82
healthcare textiles, new fibers use 1
hemicellulose 191, 192
hemodialysers 6, 36
hemodialysis membranes 44
hemostasis 106, 108, 116, 117, 145, 147
hepatitis 147, 148
hernia 95
    repair mesh 94, 95, 106, 108, 116, 117
hexokinase treated
    cellulose fibers 177
    cotton 174
        fabrics 166, 174, 175, 177, 178
        flame resistance of 178
hexokinase 162, 164, 165, 174, 179
    phosphorylation reaction 165
    treatment 163
high performance liquid chromatography
        (HPLC) 49, 50, 53, 55, 59, 61
histidine-57 27, 29
honey 12, 22, 24
human immune deficiency virus
        (HIV) 147, 148
human keratinocytes 13
human neutrophil 51
    elastase 24, 25, 51
human skin 1
hydantoinylsiloxane-treated cellolose 84
hydrogels 11, 14, 15, 24, 210
hydrogen bonding 50, 71
hydrogen bonds 4, 147, 174
hydrolase 159, 161, 178, 183, 191, 195,
        197, 198
hydrolases based enzymatic
        processes 178
hydrolysis 26, 28, 55, 87, 92, 98–101,
        107, 109, 110, 112, 161, 162,
        164, 166–175, 178, 181, 183,
        187, 188, 197

hydrophobic 6, 15, 50, 71, 86, 87, 95,
        103, 106, 129, 183, 191, 192,
        224, 225

ideal wound dressing 19, 20, 30
imide 82, 83
immunostaining 38, 39
implantable grafts 1
in vitro serum protein adsorption 36
inductively coupled plasma (ICP)
        emission spectroscopy 209–212
infection 13, 19, 20, 74, 81, 108, 109,
        112, 117, 134, 137, 146
integrins 35–38, 40
interactive biomaterials 11
interactive chronic wound dressings 21
interactive textiles 1, 7
interactive wound dressings 1, 2, 11, 21,
        22, 30
ionically derivatized cotton 23
iron 209–214
Isinglassplaster 12, 13
ivory 2

Keithley method 211
Kermel 85–88
Kevlar/PBI 85–88
kinases 159
    based enzymatic processes 178

laccases 184, 186, 191, 193–198
    textile fiber applications 195
laser 98, 99, 103, 109, 112, 113, 215
left ventricular free wall rupture 145
Lentimus edodes 126
lignin 50, 62, 63, 191–196, 198, 221
    peroxidease 196
lignocellulosic fibers 7, 209, 210, 215,
        216, 218–221, 223, 225, 226
    bonding 223
    materials 196, 218, 223, 226
    physical and chemical
        modification 215, 217
    water absorption 224
lignocellulosics 209, 210, 217, 221, 224

lignosulfate-based hydrogels 210
lipase
  concentration 170–173
  treatment 163, 178, 198
lipases 161, 163, 170, 172–174, 183,
    184, 193, 198, 199
Lister, J. 12, 14
loctite 4011 151, 152, 157
low-temperature plasma treatment 104,
    216
Lucifer Yellow (LY) 38
Lycopersicum aesculentum 129

macromolecular drugs 128, 130, 137
macrophages 18, 22, 138, 139
magnetic fibers 1
magnetic susceptibility 7, 209–214
magnetically susceptible fibers 211
magnetically susceptible paper,
    attraction 209
Mahonia aquifolium (Pursh) Nutt 127
maleic acid 221, 224
manganese peroxidase 186, 196
mannose 127
mechanical lungs 6
mechanism-based wound dressings 2, 11
  future structure and properties 11
medical textile structures, current
    technology 1
medical textiles 1, 2, 5, 24, 25, 82, 159
  extracorporeal 1
  hygienic 1
  implantable 1
  non-implantable 1
methylcyanoacrylate 153
Methylene Blue 100, 102, 110, 112,
    187, 219
methylolamide 160
microfibers 92
microwave frequencies 215
moderate temperature 178
modified cotton wound dressing fiber 5
modified fibers 1, 5, 7, 26, 28, 29, 60,
    86
  future 1

modified polyester materials 103, 105
moist wound dressings, origins 19
moisture balance 19, 30
molecular scaffolds 125
MTMIO 83, 84
  -treated cellulose 83

nanocrystalline silver-coated
    high-density polyethylene 23
nanoscale silk fibroin fibers 76
natural fibers 1, 2, 4, 7, 78, 92, 94, 104,
    181
  earliest documented evidence 2
natural polyanionic polymers 130
natural polyanions 128, 130
natural fibers, nanostructure 2, 7
natural polysaccharides 125
natural polymers 67, 200
  electrospun 6, 67
nerve toxins 1
neutral pH 178, 198, 199
n-butylcyanoacrylate 153
N-halamine 81, 83, 85–87
  in textile materials 85
  incorporation in cellulose 83
N-hydroxymethyl acryl amide 160–163,
    166, 167, 171, 178
  application 163
  cross-linked cellulose 166
Nomex 85–88
non-implantable textiles 1, 2, 5
non-implantable wound dressings 2
Nylon® 69, 85, 87, 88, 92, 95, 104,
    184–186

occlusion 11, 19, 24
octylcyanoacrylate 153, 154
octyl-2-cyanoacrylate 145
olefin 102
oligosaccharides 125, 128, 129, 198,
    200
optical micrograph 53
oxidation reactions 131, 192
oxidative enzymes 7, 182, 184, 185,
    188, 193, 197, 198

oxidized polysaccharides 125
  biological activity 125
oxidized regenerated cellulose 23
oxidoreductase 184, 191, 193, 196
oxygen permeability 30
ozone 102, 105, 215, 217

palliative treatments 12
paper 7, 69, 85, 191, 192, 194, 197–200,
    209, 216, 218, 221, 223, 225, 226
  industry 195, 197–199
  modification 193
  xylanase applications 197
papermaking process 198, 200
pathogen 1
Pavstim® 130–132, 134
peroxidases 186, 191, 193, 196
Phanerochaete chrysosporium 186
phosphorylated cotton fabric 165
phosphorylated cellulose fibers,
    dyeability 165
phosphorylated fabrics 174
phosphorylation 162, 163, 174, 179
  detection 165
  reaction 165, 166
photocatalysis supports 210
plasma 7, 99, 104–106, 109, 147,
    215–217, 224
  displays 215
  membrane 38, 41, 42, 117, 137
  treatments 71, 104, 105, 181, 182, 224
plasma-induced grafting 105
pollution control 215
polyamide (PA) 181, 182, 184–186, 188
  modification by hydrolytic
    enzymes 185
  modification by oxidative
    enzymes 185
poly(ethylene terephthalate)/PET/
    polyester 5, 69, 72, 73, 76, 5, 86,
    88, 91–107, 109–117, 159,
    182–186, 188, 217
  fabric 95, 101–103
  fibers 5, 68, 86, 93, 96, 103, 177, 183
  hydrolysis 99, 101, 183

industry 93
medical use 94
modification 91, 95, 99, 105, 106,
    183
  by hydrolytic enzymes 183
  by oxidative enzymes 184
  for medical uses 106
  in medical use, limitations 106, 107
  in routine (non-medical) use,
    limitations 95
  in routine (non-medical) use,
    modifications 95
poly(3–hydroxybutyrate) 76
poly(glycocolic) acid (PGA) 73, 74, 76
poly(lactic acid) (PLA) 73, 74, 76
poly(lactide-*co*-glycoside) 73
poly(vinyl alcohol) (PVA) 78
poly(ε-caprolactone) (PCL) 76
polyacrylonitrile (PAN) 181, 182,
    186–188
polycapronic acid 6
polycarboxylic acid 27, 160–162, 172
polycarboxylic polymers 139
  anti-fungal activity 128
  anti-tumor activity 128, 129,
    134–136, 138, 139
  anti-viral activity 128, 134, 135,
    137–139
polyester boom 92
polyglycolic acid 6
polylactic acid 6
polymer drugs 128, 137
polymer surface modification, enzymes
    for 181
polymeric additives 97
polymerization 86, 92, 96, 97, 105, 126,
    145, 146, 151, 154, 186, 192,
    194, 195
polymers 5–7, 14, 23, 67, 71, 74, 76,
    81, 82, 85, 87, 92, 93, 97, 102,
    105, 108, 109, 126, 128, 130,
    132, 139, 159, 160, 181, 182,
    186, 188, 194, 195, 200, 215,
    223, 224
surface treatments 215

polysaccharides 15, 16, 23, 125–129, 136, 138
polystyrene dishes (PS) 35, 37, 38, 40–44
polyurethane 73
  foams 15, 17, 210
porcine pancreatic elastase 52, 61
porosity 5, 49, 68, 70, 106, 108
post-surgical/wound bleeding 106, 108
potassium persulfate (PPS) 86
proliferation 6, 13, 18, 21, 35, 41, 43, 44, 116, 117, 127
prosthetics 2, 70, 71
protease
  concentration 169, 170
  sequestration 11, 28, 30, 60
  treatment 163, 166
proteases 17, 21, 23, 24, 28, 31, 42, 161, 166, 168, 193, 198, 199
protein adsorption 36, 37, 49, 50, 52, 54, 62, 63
protein binding 5, 49, 50, 52, 54, 62, 63
  to hydrolyzed polyester 110, 111
  to laser-treated polyester 112, 113
  to bifunctional polyester surface 110, 112
protein quantitation 42, 54
prototype active cotton-based wound dressings 25
pulp 191–200, 210–214, 217–226
  peroxidase applications 196
  xylanase applications 197
pulp-bleaching reactions 196

radicals 102, 104, 194, 195, 222
radiowave frequency 215, 216
reactive oxygen species (ROS) 23
rethrombosis 116, 117
RhoA, GTP-binding protein 36
Rhodococcus rhodochrous 187
root-mean-square (RMS) 219, 220
Rudbeckia 127
Rudbeckia fulgida var. *sullivanti* 127

sanitization 81
saponines 128, 129, 136
Sarcoma-180 126
SARS 81
scanning electron microscopy (SEM) 54, 63, 69, 103, 209–214, 220–222
  micrographs 103, 222, 223
sealants 6, 145, 147
Ser-195 27
sialic acids 127
Siemens, W. von 215
silent discharge CO_2 lasers 215
silk 4, 76, 77, 95, 104, 199
single wound dressing 30
site-specific drugs 128
skin substitutes 11–13, 22
smart hygienic materials 1
smart textiles 1
solvent swelling 100
spectrophotometer 53, 219
standard operational conditions 210
standard potentiometric acid titration procedures 210
Staphylococcus aureus 83–86, 88
Staphylococcus epidermidis 115
steam explosion 99, 101
  technique, to PET 102
steam-exploded PET fabrics 101, 102
sterilization 81
substrate 25, 26, 28, 29, 31, 40, 49–52, 55, 62, 97, 104, 109, 160, 161, 168–172, 174–176, 178, 183–185, 188, 193–197, 211, 212, 214, 217, 218
sulfonated polysaccharides 128
surface
  carboxylic acids 210, 218
  modification 1, 71, 105, 159, 181, 185, 188, 191, 193, 215
  polymer additives 97
  resistivity 209, 210, 214
  treatments 98, 182, 215–221, 223–226

Swiss 3T3 murine fibroblast 36, 41
synthetic fibers 4, 7, 85, 86, 91, 103,
    188
synthetic membranes 44
synthetic polymers 6, 23, 67, 85, 92,
    159, 181, 223
    electrospun 6
synthetic textiles 181, 182, 188

talin 37
TAPPI standard method 210
tensile energy absorption 213
tensile strength 2, 69, 149, 150, 152,
    154, 161, 164, 167, 169–173,
    178, 195, 209, 210, 213, 214,
    224
Terylene™ 92
textile
    material 1, 82, 85, 160, 163, 164, 166,
        179
    products 7, 191, 194, 200
    substrates, glucose oxidase
        bleaching 197
Thamnalia subuliformis 127
Thamnolan 127
Thermomonspora fusca 184
thermoplastic extrusions 210
thrombosis 106, 108, 111
time-release drug polymers 128
time-to-change indicator 21
tissue
    adhesives 6, 145, 146, 153
    encapsulation 5
    engineering 6, 21, 45, 71, 73, 76, 78
    repair 3, 4, 95
    sealing 4, 145
titanium IV 92
tomatoside 129, 131, 132, 134
total protein cell content 41
transcytosis 137
transmission electron microscopy 38
treatment
    corona 102, 103
    excimer laser 103
    flame 103, 105

laser 98, 99, 112
ozone 105
palliative 12
plasma 71, 104, 105, 181, 182, 224
segment 113–115
UV 105
wound 12, 19
triallyl-1,3,5-triazine-2,4,6(1H,3H,5H)-
    trione (TATAT) 86, 87
tris[hydroxymethyl]aminomethane
    (trizma) 52, 54
Trypticase Soy Agar (TSA) 115

unmodified polyester 103
UV 53, 59, 165
    excimer lamps 98, 215
    irradiation 103
    treatments 105

van der Waals 50
vascular grafts 5, 94, 95, 106, 107, 109,
    117
vegetable fibers 12
vinculin 37, 38
VEGF 117
virus transmission 147
vitamin E modified cellulose 6
vitronectin (VN) 35, 36, 40, 109
volume resistivity 213, 214

wood fibers 215
wooden dentures 2, 7
wool 4, 102, 104, 196
    fibers 199
wool-based dressings 13
wound care products 21, 29, 30
wound dressings 1–3, 17–24, 26, 27, 29,
    30, 49, 51, 52, 60, 71, 78, 94,
    106, 108
    active cotton-based, prototype design
        of 25, 26
    carbohydrate-based 11, 22, 24
        assessment 24
        design 24
        preparation 24

wound dressings (*cont.*)
  current developments 19
  fibers 5, 12, 18, 26, 49, 52
    design 52
    historical characteristics 12
  future 18
  ideal 19, 20, 30
  interactive 1, 2, 21, 22, 30
    chronic 21, 26, 29, 30
  materials 12, 14, 19, 22
  mechanism-based 11
  moist, origins of 19
  non-implantable 2
  occlusive 12, 20, 60
  prototype active cotton-based
    25

wound healing 2, 5, 6, 11–13, 18–20,
  22, 23, 51, 52, 60, 63, 67, 116,
  117, 147
  science of 13
wound occlusion 19
wound proteases, sequestration 23
wound protein binding 49
wound treatments 19
woven medical textile materials 1
Wrinkle recovery angle (WRA) 164,
  167, 169–173

xerogels 11, 24
XPS 174, 187, 188
xylanase 197–199
xyloglucans 200